Relativistic Flight Mechanics and Space Travel
A Primer for Students, Engineers and Scientists

Richard F. Tinder
Professor Emeritus of Electrical Engineering and Computer Science,
Washington State University

SYNTHESIS LECTURES ON ENGINEERING #1

MORGAN & CLAYPOOL PUBLISHERS

ABSTRACT

Relativistic Flight Mechanics and Space Travel is about the fascinating prospect of future human space travel. Its purpose is to demonstrate that such ventures may not be as difficult as one might believe and are certainly not impossible. The foundations for relativistic flight mechanics are provided in a clear and instructive manner by using well established principles which are used to explore space flight possibilities within and beyond our galaxy.

The main substance of the book begins with a background review of Einstein's Special Theory of Relativity as it pertains to relativistic flight mechanics and space travel. The book explores the dynamics and kinematics of relativistic space flight from the point of view of the astronauts in the spacecraft and compares these with those observed by earth's scientists and engineers—differences that are quite surprising.

A quasi historical treatment leads quite naturally into the central subject areas of the book where attention is focused on various issues not ordinarily covered by such treatment. To accomplish this, numerous simple thought experiments are used to bring rather complicated subject matter down to a level easily understood by most readers with an engineering or science background. The primary subjects regarding photon rocketry and space travel are covered in some depth and include a flight plan together with numerous calculations represented in graphical form. A geometric treatment of relativistic effects by using Minkowski diagrams is included for completeness. The book concludes with brief discussions of other prospective, even exotic, transport systems for relativistic space travel. A glossary and simple end-of-chapter problems with answers enhance the learning process.

KEYWORDS

Rocket, Photon, Flight, Relativity, Mechanics, Space, Astronauts, Exploration, Star, Galaxy, Einstein, Minkowski.

Contents

Preface viii

Glossary of Terms and Expressions xi

Glossary of Symbols and Abbreviations xxi

1. Introduction 01
   1.1 Scope and Philosophy of this Book 01
   1.2 Is Interstellar Space Travel Possible? 01
   1.3 Why Travel to Distant Stars, Planetary Systems, and Galaxies? 02
   1.4 General Requirements for Interstellar Space Travel 03

2. Background 05
   2.1 The Nature of Light and its Velocity 05
   2.2 Michelson–Morley Experiment and the Null Result 07
   2.3 Postulates of Einstein's Special Theory of Relativity 10
   2.4 Principle of Simultaneity and Synchronicity 11
   2.5 Relativistic Time Dilation and the Clock Paradox 13
   2.6 Relativistic Distance Contraction 17
   2.7 Relativistic Transformation of Coordinates and Addition of Velocities . 20
   2.8 Relativistic Momentum and Mass 25
   2.9 The Relativistic Mass–Energy Relation 27
   2.10 Problems 30

3. Relativistic Rocket Mechanics 33
   3.1 Relativistic (Proper) Measurements and Calculations of the Astronauts . 33
   3.2 Introduction to Rocket Mechanics; the Rocket Equation 35
   3.3 The Photon Rocket 39
   3.4 Relativity of Velocity, Time, Acceleration, and Distance 44
   3.5 Energy Requirements for Relativistic Flight 54
   3.6 Concluding Remarks Regarding Relativistic Flight 56
   3.7 Problems 57

CONTENTS

4. Space Travel and the Photon Rocket ... 59
- 4.1 Summary of Important Equations ... 59
- 4.2 The Flight Plan and Simplifying Assumptions ... 60
- 4.3 Comparative Distances, Flight Times and Rocket Mass Ratios to Various Star/Planet Systems, Galaxies and Beyond—Calculation Plan ... 62
  - 4.3.1 Star/Planet Systems, Galaxies, and Beyond ... 64
- 4.4 Consequences of Particulate Efflux Velocities, Varying Speeds of Light, and Acceleration ... 70
  - 4.4.1 Consequences of Particulate Propulsion Systems ... 70
  - 4.4.2 Consequences of a Varying Speed of Light ... 71
  - 4.4.3 Consequences of a Varying Proper Acceleration ... 74
  - 4.4.4 Consequences of Combining Variable Light Speed with Variable Acceleration ... 76
- 4.5 "Give me *WARP* Three, Scotty"; Practical Considerations ... 76
  - 4.5.1 Efficiency Considerations ... 77
  - 4.5.2 Payload, Engine, Crew and Reactor Hardware Mass Considerations ... 77
  - 4.5.3 Energy Requirements ... 78
  - 4.5.4 Life-Preserving and Health Issues of the Astronauts ... 80
- 4.6 Problems ... 81

5. Minkowski Diagrams, K-Calculus, and Relativistic Effects ... 83
- 5.1 Minkowski Diagrams ... 83
- 5.2 K-Calculus and Relativistic Effects and Measurements ... 85
  - 5.2.1 Time Dilation ... 85
  - 5.2.2 Relativistic Doppler Effect ... 87
  - 5.2.3 Redshift—Velocity and Distance Determinations ... 88
  - 5.2.4 Hubble's Law ... 90
  - 5.2.5 Parallax Method—Distance Measurements ... 91
  - 5.2.6 Composition of Velocities ... 91
- 5.3 Problems ... 93

6. Other Prospective Transport Systems for Relativistic Space Travel ... 95
- 6.1 Nuclear Particle Propulsion ... 95
- 6.2 Matter/Antimatter Propulsion ... 96
- 6.3 Laser Sail Propulsion ... 96
- 6.4 Fusion Ramjet Propulsion ... 97
- 6.5 Exotic Space Transport and Propulsion Systems ... 97

6.5.1		Gravity-Controlled Transport	98
6.5.2		Transport in Variable Light Speed Media	98
6.5.3		Wormhole Transport and Time Travel	99
6.5.4		Warp Drive Transport	99

Appendix A Fundamental Constants, Useful Data, and Unit Conversion Tables 101
A.1 Fundamental Constants and Useful Data 101
A.2 Units of Conversion 101
A.3 Metric (SI) Multipliers 103

Appendix B Mathematical Definitions and Identities 105
B.1 Hyperbolic Functions 105
B.2 Logarithm Identities (Base b) 105

Appendix C Derivation of The Rocket Equations 107
C.1 The Photon Rocket Equation 107
C.2 The Effect of Efficiency 108
C.3 The Classical Rocket Equation 108

Endnotes 111

Index 113

Preface

By means of present-day transport systems, we can travel into interstellar and intergalactic space and back within the lifetimes of the astronauts.

This statement is nonsense, of course! The purpose of the book is not to detail the means by which interstellar and intergalactic space travel can be carried out now or even in the near future. What we purport to do is to demonstrate that such ventures are not impossible but require knowledge of the universe yet to be gained and the existence of technologies yet to be discovered and developed. We have to remember that our current understanding of the physical universe is perhaps only 100, 200, 300, or more years old. You pick! In the past few decades, the computer/space-age has yielded surprising and fantastic revelations of the visible sky and beyond into the far reaches of the universe nearly 14 billion light years from the earth. With only theories and models and some experimental work to go by, we bandy about expressions such as the big bang, expanding universe, black holes, wormholes, relativity, dark matter, dark energy, multiverses, warp drives, the effect of gravity on the fabric of spacetime, superstring theory, M-theory and branes, and a host of other interesting but usually esoteric topics. But what do we really know for certain? Answer: Actually, not much!

It is surprising how any deviation from an established norm can rankle some well-established individuals in certain fields such as astronomy, cosmology, and rocket science. Please believe that it is not the aim of this book to do that. Though it may inadvertently happen, it is really our intent to lay out the foundation for relativistic flight mechanics by using well-established principles. We will also move into fields not well established, as we must to accomplish the primary goal of the book: to explore the possibility of space travel that depends on future technological advancements and a much greater understanding of the physical universe. Our theme is: *the travel to stars and distant galaxies may not be as difficult as you may believe and certainly not impossible.* The rate at which information is obtained via solar-system space exploration, advanced ground-based telescopes, earth-orbiting telescopes, earth-trailing telescopes orbiting the sun, multiple orbiting telescope interferometer systems, and orbiting x-ray and microwave telescopes, assures us of dramatic and surprising results in the very near future. But that must be only the beginning. Until we can develop the appropriate advanced transport systems needed for space travel, and until we can explore and conduct experiments outside of the solar system, we will be unable to venture into interstellar and intergalactic space.

The main substance of the book begins with a background review of Einstein's Special Theory of Relativity as it pertains to relativistic flight mechanics and space travel. Next, the book moves into relativistic rocket mechanics and related subject matter. Finally, the primary subjects regarding space travel are covered in some depth—a crescendo for the book. This is followed by a geometric treatment of relativistic effects by using Minkowski diagrams and K-calculus. The book concludes with brief discussions of other prospective, even exotic, transport systems for relativistic space travel. Overall, this book is an extension of the work of Eugen Sänger, a pioneer in relativistic space flight mechanics that provides an encouraging backdrop for this book.

An appendix is provided to cover tables of useful data and unit conversions together with mathematical identities and other information used in this book. Endnotes are provided for further reading. A detailed glossary and index are given at the beginning and end of the book, respectively. To provide a better understanding of the subject matter presented in the book, simple problems with answers are provided at the end of each of the four substantive chapters. Problem solutions are available to course instructors for use with the text.

Acknowledgements

The author would like to acknowledge the valued contributions of emeritus professors of mechanical engineering Drs. Clayton T. Crowe and Timothy R. Troutt for their critical insight into some of the more complex subject matter presented in this book. Their many discussions, arguments and pontifications with the author have helped to make this book a valuable resource of information on what can only be described as unusual but fascinating subject matter.

These acknowledgements would not be complete without recognizing the encouragement of and helpful conversations with Joel Claypool, publisher of Morgan & Claypool Publishers. Then finally and most importantly, the author wishes to express his gratitude for the support of his loving wife, his best friend and confidant, Gloria.

Richard F. Tinder
Pullman, Washington

Glossary of Terms and Expressions

Absolute zero. The lowest known temperature at which a substance is void of heat energy; 0° K or −273 °C.

Acceleration. A time rate of change in velocity.

Accelerometer. A device used to measure acceleration.

Adiabatic process. A condition in which a process exchanges no heat with its surroundings.

Antigravity. The concept of reducing or reversing the force of gravity.

Antimatter. Matter that has the same gravitational properties as ordinary matter but is of opposite charge including opposite nuclear force charges.

Antiparticle. A particle of antimatter.

Atom. The fundamental building block of matter, consisting of a nucleus (made up of protons and neutrons) and an orbiting swarm (probability cloud) of electrons.

Black hole. A region in spacetime from which nothing can escape not even light because the gravitational forces are so strong.

Blueshift. The bluing of light radiation emitted by an object that is approaching an observer, caused by the relativistic Doppler effect.

Bondi, Herman. Astrophysicist known for his earlier work interpreting relativity principles.

Burnout. The conclusion of a period of thrust operation, usually at fuel exhaustion.

Boundary conditions. The state of a system at a boundary in time and space.

Calculation plan. Knowing the distance to a destination star or galaxy and the proper acceleration of the spacecraft, the proper relativistic quantities can be calculated.

Characteristic value. The value of any quantity that is measured or calculated by astronauts in a spaceship traveling at relativistic velocities; also eigenvalue or proper value.

Charge. The property of a particle (positive or negative charge) that determines how it responds to an electromagnetic force.

Classical theory. Theory based on concepts established before the advent of relativity and quantum mechanics; also Galilean or Newtonian theories.

Clock (twins) paradox. Comparative times elapsing between a stay-at-home observer to that of his/her astronaut twin returning from a space flight at relativistic velocities.

Conservation of energy. The law of science that states that energy (or its mass equivalent) can neither be created nor destroyed.

Conservation of momentum. The law of mechanics that requires the total momentum before and after a collision event to be equal.

Correspondence Principle. As applied in this text—the principle requiring that relativistic mechanics and classical mechanics must correspond where their realms of validity overlap.

Cosmic horizon. Term applied to the limit of the detectable universe presently taken to be about 14 billion light years away from us.

Cosmic strings. Extremely dense thin threads of matter and energy theorized to crisscross the universe as vestiges of the "big bang"—strings near which the velocity of light may greatly exceed the present known value of c.

Cosmology. The study of the universe as a whole.

Cosmos. The universe and all that it encompasses.

Dark Energy. Energy within the universe thought to be associated with dark matter and thought to drive the increasing expansion of the visible universe.

Dark matter. Matter comprising as much as 90% of the matter in the universe, thought to exist within and between galaxies and star clusters, and that cannot be observed directly but which can be detected by its gravitational effects.

Deuterium. An isotope of hydrogen the nucleus of which consists of a neutron and a proton.

Doppler effect. The frequency shift in sound or light waves perceived by an observer when the emitting source is in motion relative to the observer.

Doppler redshift. Means by which the Doppler effect is used to measure the velocity of recession of distant galaxies.

Dynamics. The mechanics of motion independent of mass.

Einstein number. The ratio of a spacecraft's velocity v, as determined from the earth, to the velocity of light, c.

Efflux. That which is expelled from the plane of the nozzle(s) of a spacecraft or rocket.

Efflux power. Kinetic power of the efflux; thrust × efflux velocity.

Electromagnetic radiation. The radiant energy carried by an electromagnetic wave.

Electromagnetic wave. A wavelike disturbance in an electromagnetic field that travels at the speed of light and that may include a range of wavelengths from radio waves to gamma-ray waves, a range of 15 orders of magnitude.

Electron. A negatively charged particle that typically orbits the nucleus of an atom.

Elementary particle. A particle that is believe not to be subdivided into other particles.

Escape velocity. The velocity required for a vehicle to overcome a gravitational field.

Ether. A hypothetical nonmaterial medium once presumed to occupy all space and be responsible for the propagation of electromagnetic radiation. It is now believed not to exist.

Ether wind. A wind thought to exist due to the earth's movement through space as it orbits the sun at a velocity of about 30 km s^{-1}.

Event. A point in spacetime specified by its place and time. See Minkowski diagram.

Event horizon. The surface around a black hole within which nothing can escape the powerful gravitational forces, not even light.

Field. Something that exists throughout space and time.

Fission. The separation of a nucleus into two nuclei and a number of fundamental particles with the simultaneous release of energy.

Flight plan. A constant acceleration to midpoint between the earth and a destination star or galaxy followed by a constant deceleration over the remaining distance to the destination.

Force. The time rate of change of momentum.

Force field. The means by which a force communicates its influence.

Free space. Space completely free of fields and not acted upon by any force.

Frequency. The number of complete wave cycles a wave completes each second.

Frozen flow efficiency. Efficiency that accounts for certain reaction mass losses during the generation of a propulsive efflux.

Fusion. The combination of two nuclei into a single nucleus with the release of energy.

Fusion ramjet propulsion. A propulsion system employing a huge magnetic funnel to scoop up interstellar hydrogen into a reactor as fuel where fusion takes place to produce propulsive power.

Galaxy. An astronomically large cluster of stars and other galactic material the center of which may contain a black hole.

Galilean transformation equations. Classical equations for transforming coordinates in four-dimensional space where time and space are considered to be independent.

Gamma ray. Electromagnetic radiation of wavelengths in the range 10^{-10}–10^{-13} m produced by nuclear reactions.

General relativity. Einstein's formulation of gravity showing that space and time communicate the gravitational force through the curvature of a four-dimensional spacetime continuum.

Gravitational field. The means by which gravity exerts its influence.

Gravitational force. The weakest of the fundamental forces of nature.

Gravitational wave. A wavelike disturbance in a gravitational field.

Graviton. The messenger particle of zero mass thought to be associated with a gravitational force.

Hertz. Frequency unit (Hz) equal to 1 cycle per second.

Hubble parameter. The constant of proportionality between velocity and distance defined in Hubble's law.

Hubble's law. The law that states that the redshift (hence also velocity) of receding galaxies is linearly proportional to their distance from the earth.

Imaginary time. Time measured by using imaginary numbers as, for example, $i\tau$ where $i = \sqrt{-1}$.

Impulse. Momentum change; thrust delivered over a time interval.

Internal efficiency. It lumps together all the efficiency factors (thrust efficiency, combustion efficiency, frozen flow efficiency, etc.) that influence the production of a directed particulate efflux in the rocket's engine or reactor chamber.

Inertial frame of reference. A frame of reference that is at rest or of uniform (nonaccelerating) motion along a straight path; a frame that obeys Newton's first law of motion.

GLOSSARY OF TERMS AND EXPRESSIONS

Infinity. A boundless or endless extent. Any number divided by zero.

K-calculus. The mathematics associated with Minkowski diagrams.

Kelvin. A temperature scale in which temperatures are given relative to absolute zero.

Kinematics. Mechanics involving mass as it relates to motion.

Kinetic power of rocket. Thrust × velocity of the rocket.

Laser. For "light amplification by stimulated emission," the laser produces a very intense, highly directional, beam of light which is monochromatic and coherent (in-phase radiation).

Laser sail propulsion. A propulsion system involving a light collector to capture light from the sun, a laser, and a mirror, all as needed to cause the laser beam to propel a sail/spacecraft in space.

Light. As used in this book, refers to any electromagnetic radiation.

Light cone. A surface in spacetime within which are defined the possible directions of light rays that can pass to or through a given event.

Light year. The distance traveled by light in one year, equal to 3.0×10^8 m s^{-1}.

Lorentz contraction. A feature of special relativity in which moving objects appear shortened as viewed along their direction of motion; also Fitzgerald contraction.

Lorentz transformation equations. Transformation equations required to derive the composition of relativistic velocities as opposed to the classical or Newtonian sum of velocities.

Lyman break. Redshifts produced by a drop-off in the ultraviolet spectrum of hydrogen from some galaxies called Lyman galaxies.

Mach number. Used to express supersonic flight velocities (e.g., mach 1, mach 2, etc.).

Macroscopic. Scales of observation typically encountered in our everyday world; roughly the opposite of microscopic.

Magnetic field. The field responsible for magnetic forces.

Mass. The matter in a body that resists acceleration in free space; a measure of a body's inertia.

Mass ratio. The ratio of the initial mass of a vehicle to its instantaneous mass (e.g., at burnout).

Matter. All substances are composed of atoms, which in turn are made up of quarks and electrons.

Messenger particle. The smallest bundle of a field force responsible for conveying a force.

Messier, Charles. Astronomer for whom many galaxies are named and categorized (e.g., M31).

Microscopic. Scales of observation less than what can be seen by the naked eye, typically less than 0.01 mm.

Minkowski diagram. A geometrical interpretation of special relativity within the light cone.

Momentum. In kinematics, the product of mass and velocity.

Momentum thrust. The component of thrust provided by the change in propellant momentum.

Motion. The movement of an object that may or may not contain mass.

Multiverse. The hypothetical enlargement of the cosmos in which our universe is but one of the extremely large number of other separate and distinct universes.

Muon. Fundamental particle of short life and having the mass of about one-tenth that of a proton.

Negative matter. Mass with gravitational properties opposite to that of ordinary matter.

Neutrino. Chargeless species of particle subject only to the weak force.

Neutron. Chargeless particle, typically found in the nucleus of an atom consisting of three quarks.

Newton's first law of motion. Law that states, "Every body continues in its state of rest or uniform motion in a straight line unless required to change that state due to forces acting on it."

Newton's second law of motion. Law that states, "Acceleration of an object is proportional to the force acting on it and inversely proportional to its mass."

Newton's third law. Law that states, "Every action is associated with an equal and opposite reaction."

Noninertial frame of reference. A frame of reference that is accelerating or is being acted on by external forces. A reference frame in which Newton's first law does not hold.

Nuclear fission. The process by which a nucleus breaks down into two or more smaller nuclei with the accompanying release of energy.

Nuclear fusion. Process by which two nuclei collide and join to form a larger, heavier nucleus.

GLOSSARY OF TERMS AND EXPRESSIONS xvii

Nuclear propulsion. A propulsive system for which the efflux consists of nuclear particles and electromagnetic radiation resulting from a fission or fusion reactor.

Nucleus. The core of an atom consisting of protons and neutrons.

Observer. As used in this book, a fictitious person placed in a reference frame or on a line in a Minkowski diagram for the purpose of emitting or receiving an electromagnetic radiation signals as, for example, light.

Parallax method. A heliocentric method of measuring distance of a star or galaxy from the earth, given in parsecs. Use is made of the radius of the earth's orbit around the sun to determine the greatest parallax (angle) in the star's directions from the earth and sun during a year. A small correction is used to bring the earth to its average distance from the sun.

Parsec. A measure of distance as determined by the parallax method.

Particulate propulsion. A propulsive system for which the efflux consists of particles as, for example, chemical propulsion.

Payload. Mass of a space vehicle in excess of what is required to propel and operate the vehicle.

Phase velocity. The velocity of light in a liquid equal to the velocity of light in a vacuum divided by the refractive index of the liquid.

Photoelectric effect. Means by which electrons are ejected from a metallic surface when light falls on it.

Photon. The smallest bundle or packet of the electromagnetic force field; messenger particle of electromagnetic force.

Photonic propulsion. A propulsive system for which the efflux consists of photons.

Plank length. The scale below which quantum fluctuations in the fabric of spacetime become enormous—about 10^{-33} cm.

Planck's constant. Denoted by symbol $\hbar$, Planck's constant is a fundamental parameter that determines the size of the discrete units of energy, mass, etc. into which the microscopic world is partitioned.

Planet. A major celestial body orbiting a star.

Positron. The positively charged antiparticle of the electron.

Postulates of special relativity. Core principles of special relativity declaring that the laws of science are valid in all inertial reference frames, that all inertial frames are equivalent, and that light propagates through matter-free space at a constant velocity independent of the state of motion of the emitting source or observer.

Propellant. The working efflux from a propulsive system that provides the thrust resulting from the action of Newton's third law.

Proper Einstein number. The ratio of a spacecraft's proper velocity v_e (as calculated by the astronauts) to the speed of light, c.

Proper quantity. Any quantity that is measured or calculated by the astronauts in a spacecraft; including proper time, acceleration, velocity, distance, mass, energy, etc.; also eigenvalue or characteristic value.

Propulsive efficiency. Efficiency of a propulsive system that accounts for the residual power lost by the ejected efflux.

Proton. A positively charged particle found in the nucleus of an atom, consisting of three quarks.

Quantum (pl. quanta). The smallest physical unit into which something can be partitioned. For example, photons are the quanta of the electromagnetic field.

Quark. A charged fundamental particle that exists in six varieties, each having a different mass.

Radiation. The energy carried by waves or particles through space or some other medium.

Reaction mass. Fuel mass used to produce a propulsive thrust.

Redshift. The reddening of light radiation emitted by an object that is receding from an observer, caused by the relativistic Doppler effect.

Relativistic Doppler effect. The shift in the frequency of light emitted by a moving celestial body relative to an observer.

Relativistic effects. The effects that relativistic velocities have on time, distance, mass, momentum, and energy, and all the quantities that are derived there from. Examples include time dilation and distance contraction.

Relativistic mass. A misnomer invoked to characterize photon momentum and thrust.

Relativistic transformation equations. The transformation equations that lead to the Lorentz transformation equations from which the composition of velocities is derived.

Rest energy. The energy of an object at rest.

Rest frame. A frame of reference at rest relative to an inertial or noninertial frame in motion.

Rest mass. The mass of an object at rest relative to a given rest frame.

Rocket. As used in this book, a space vehicle having a particulate or photonic propulsion system.

Rocket equation. Equation that relates proper Einstein number to the proper mass ratio.

Singularity. A point in spacetime at which the spacetime curvature becomes infinite.

Spacetime. Union of space and time originally emerging from special relativity that constitutes the dynamical four-dimensional arena within which the events of the universe take place.

Special relativity. Einstein's theory based on the ideas that the laws of science are valid in all inertial reference frames, that all inertial frames are equivalent, and that light propagates through matter-free space at a constant velocity independent of the state of motion of the emitting source or observer.

Specific energy. The instantaneous proper energy of the rocket per unit proper mass of the rocket.

Specific impulse. The effectiveness of a propellant expressed in terms of thrust per unit mass consumption rate of propellant; generally equal to the proper efflux velocity.

Star. A discrete self-luminous body in space for which the process of fusion is the likely cause of the luminescence.

Superluminal. It designates velocities in excess of the velocity of light, c.

Tachyons. Hypothetical particles that must travel faster than the speed of light and that require energy to slow them down toward the speed of light. Special relativity would require that tachyons move backward in time. Tachyons have never been observed experimentally.

Thermal efficiency. Ratio of observed total enthalpy difference to the theoretical total difference.

Thrust. The propulsive reactive force produced by the action of Newton's third law in producing the propulsive efflux from a space vehicle; time rate of change of mass × efflux velocity.

Thrust efficiency. Ratio of the actual to theoretical values for thrust to efflux power ratio in percentage.

Time dilation. A feature emerging from special relativity in which all mechanical and biological clocks slow down for an observer traveling at relativistic velocities relative to an observer in a rest frame.

Tritium. A rare radioisotope of hydrogen having an atomic mass of 3.

Tsiolkovsky rocket equation. The classic rocket equation that relates the velocity of a rocket to its efflux velocity and initial-to-instantaneous mass ratio.

Uniform velocity. Constant velocity with no inertial effects; also uniform motion.

Vacuum. The absence of all known matter in space—empty space.

Velocity. The relative speed and direction of a body in motion.

Warp drive transport. Space drive based on the idea of expanding space behind a space vehicle and contracting space ahead of it thereby producing a manipulation of spacetime.

Warp number. A means of designating the proper Einstein number (as calculated by the astronauts)—similar to the mach number used to express supersonic flight velocities.

Wavelength. The distance between successive peaks or troughs of a waveform.

Weight. The force exerted by gravity on a body.

Wormhole. A tube-like region in the spacetime continuum connecting one region of the universe to another. Wormholes may also connect parallel universes in a multiverse and may provide the possibility of time travel.

Glossary of Symbols and Abbreviations

AU	astronomical unit equal to the mean distance from the earth to the sun
b	relative acceleration of an object as determined by observers on earth
b_e	proper acceleration as measured by astronauts
c	velocity of light
E	energy
E_K	kinetic energy
F_e	proper thrust produced by a propellant
f	frequency; final limit of integration
f_O	frequency of light perceived by an observer
f_S	frequency of light emitted by a moving source
H_O	Hubble parameter
I_{SP}	specific impulse
i	used to indicate an imaginary number, $i = \sqrt{-1}$; initial limit of integration
$i\tau$	imaginary time as used in Minkowski diagrams where $\tau = ic$
K	ratio of the time intervals between successive light pulses received by an observer in motion to those sent by an observer at rest—the K-factor used in Minkowski diagrams
kpc	kiloparsec
M	reaction fuel mass of a rocket
m	mass
m_e	instantaneous proper mass of a rocket; rocket mass at burnout
m_{eo}	initial proper mass of a rocket
m_{eo}/m_e	the mass ratio as defined in the rocket equation
m_0	rest mass of an object
$m_0 c^2$	rest energy of an object
n	index of refraction of a liquid

xxii GLOSSARY OF SYMBOLS AND ABBREVIATIONS

P	mass of payload, propulsion engine hardware, reactor hardware and crew
p	momentum
pc	parsec
P_e	proper efflux power
R	a frame of reference
S	distance from the earth to a spacecraft, star or galaxy as measured by observers on earth
S_e	instantaneous proper distance as calculated by astronauts
S_e^*	distance from spacecraft to earth as measured by astronauts
T	time of travel as determined by observers on earth; imaginary axis of a Minkowski diagram
T_e	proper time of travel to a destination as observed by astronauts
t	time as measured by observers on earth
t_e	proper time as measured by astronauts
u_e	proper efflux velocity of a propellant
v	relative velocity as calculated by observers on earth
v_e	instantaneous proper velocity as calculated by astronauts
v/c	Einstein number as obtained by observers on earth
v_e/c	proper Einstein number as calculated by astronauts; pseudorotation in spacetime
$(v_e/c)_{Avg}$	average proper Einstein number
z	an axis in Cartesian coordinates

Greek Symbols

ε	specific energy (energy per unit mass)
ε_0	permittivity in a vacuum
η_I	internal efficiency of the fuel reactants to form a properly directed efflux
η_P	propulsive efficiency
η_T	total efficiency $= \eta_I \cdot \eta_P$
λ	wavelength
μ_0	permeability in a vacuum
τ	time as used in a Minkowski diagram

CHAPTER 1

Introduction

1.1 SCOPE AND PHILOSOPHY OF THIS BOOK

It is the intent of this book to introduce the reader to the rocket mechanics principles as applied to relativistic flight. At the same time, we will develop the prospects for space travel by assuming the use of technologies that are not presently available but which may be available in the future. We need only look back some 50 to 75 years to be awestruck at the technological developments that have become commonplace today. Solar system space exploration, the computer age, genetic engineering, nanotechnology, deep space astronomy, and a host of other remarkable developments—these set the stage for other more fascinating and remarkable discoveries to come. Remember: "A tool begets a tool, begets a tool, etc." We are in a geometrical if not an exponential slope of technological development that portends an amazing and mind-boggling future. Imagine the far-reaching consequences of just a few developments that could dramatically impact technology, our lives, and in particular affect space exploration. These developments may include room temperature superconducting materials, controlled fusion, production and containment of antimatter, understanding and control of gravity, development of quantum computers, knowledge and use of dark matter and dark energy believed to occupy interstellar space, and so on and so forth.

We will explore the dynamics and kinematics of relativistic space flight from the point of view of the astronauts in the spacecraft and compare these with those observed by earth scientists and engineers. A quasihistorical treatment leads quite naturally into the central subject areas of the book, and every effort is made to bring rather complicated subject matter down to a level easily understood by most readers with an engineering or science background. To accomplish this, use is made of numerous simple thought experiments designed to facilitate understanding of complex concepts.

1.2 IS INTERSTELLAR SPACE TRAVEL POSSIBLE?

Perhaps a better question would be: Is interstellar space travel impossible? An honest answer would be: No, it is not impossible but interstellar space travel to even the nearest star cannot be done with present-day technology, at least not in any reasonable period of time. Of course, we can send a spacecraft out beyond the solar system but at such low velocities that the transit time to our nearest star would be prohibitive. For example, a spaceship traveling at 25,000 miles per hour (mi h^{-1}) (approximately the escape velocity from the earth) would require 114,000 years to reach the nearest star, alpha centauri, at a distance of 4.3 light years (Lt Yr) or 2.52×10^{13} miles from the earth. Clearly, this fact would not inspire any astronaut to make the journey much less turn on the political or private machinery required to fund such a venture. But what if advanced technology were available, technology that would allow a spacecraft with astronauts to be propelled into deep space at relativistic speeds, meaning at some fraction of the speed of light. Think of the impact that it would have on every aspect of our existence here on earth. Can we rule out the possibility that new discoveries will yield a technology that makes interstellar space travel possible? We will consider the theoretical problems involved and attempt to show that interstellar space travel is possible but very much dependent on new future technological developments.

1.3   WHY TRAVEL TO DISTANT STARS, PLANETARY SYSTEMS, AND GALAXIES?

There are more than 160 planets associated with more than 100 star/planetary systems that have been discovered over the past few years. The discovered planets are generally several times the size of our Jupiter and are associated with stars that are tens to hundreds of light years from the earth and of magnitude that cannot be seen by the naked eye. Planets of the size of earth have not been discovered because our telescopes are not yet powerful enough to detect their presence. But that will change in the very near future when it is expected that many striking revelations will be made. For example, there are several post-Hubble telescopes and orbiting laser interferometers in the making that will most certainly expose many secrets of the universe that have escaped our knowledge and understanding up to now. It is likely that one or more of these telescopes will be able to detect the presence of earth-like planets that are associated with stars in our galaxy. They may provide vitally important information regarding the presence and nature of dark matter, and the dark energy thought to be responsible for the accelerating expansion of the universe. The discovery of galactic antimatter and the possibility of existing antigravity are the findings that may also lie ahead in the coming years and decades. Our knowledge of the universe is so limited and incomplete that it begs for enlightenment; enlightenment that could come at any time with an astounding influence on our perception of life and its existence in the universe.

So why travel to distance stars, planetary systems and galaxies? Think what it would mean to find other life forms outside of our solar system. Think of what we could learn from intelligent beings that may be thousands (millions?) of years more advanced than we are. Think of the many questions we have regarding our existence that more advanced beings could answer. The impact these possibilities would have on every aspect of our life on earth is so profound that it defies any attempt at a reasonable response. While we have none of the answers to these questions, we will examine the many factors that would make relativistic flight to distance stars and galaxies possible within the life-time of the astronauts.

1.4 GENERAL REQUIREMENTS FOR INTERSTELLAR SPACE TRAVEL

Putting it simply, the problem with present-day rocketry is the use of rocket-borne propulsive fuel systems that generate little more than enough thrust just to move the fuel mass. This means that the initial to final fuel-mass ratio must necessarily dwarf any payload present. While such conventional propulsive (chemical) systems have been used successfully to place satellites in orbit and to send men to the moon, etc., they are of no use for interstellar space travel. So what are the requirements of successful interstellar space travel? The following is a list of the salient requirements not necessarily related:

1. Any rocket-borne propellant fuel system must convert fuel to useful propulsive power at high efficiency, ideally at nearly 100% efficiency.
2. The propellant (efflux) velocity must be near or at the speed of light (photon propulsion).
3. The propulsive or transport system must be able to accelerate the space vehicle to relativistic velocities, i.e., to some fraction of the speed of light.
4. New transport systems are needed which rely less on the need for a propellant.
5. New methods of onboard energy production are required for these advanced systems.

These and other requirements will be the subject of discussion at various places in the following chapters.

CHAPTER 2

Background

2.1 THE NATURE OF LIGHT AND ITS VELOCITY

Since the velocity of light plays an essential role in the development of relativistic flight mechanics, it is appropriate that we discuss its nature, its behavior, and its velocity in certain media. To start with, visible light ranges from violet to red in the wavelength range of 400 nm to 750 nm, respectively (1 nm = 10^{-9} m). Wavelengths shorter than 400 nm are called ultraviolet (UV), and wavelengths greater than 750 nm are called infrared (IR). If the wavelengths are given in angstroms (1 Å = 10^{-10} m), then visible light would fall in the range of 4000 Å to 7500 Å. In any case, all light spectra from UV to IR are part of a much larger electromagnetic spectrum which ranges from radio waves at about 3×10^3 m wavelength to gamma rays at about 3×10^{-12} m. Clearly, the visible light range occupies only a tiny portion of the electromagnetic spectrum. IR, microwaves, and radio waves have wavelengths that are much larger than those of the visible spectrum while UV, x-rays, and gamma rays have wavelengths much shorter than those of the visible spectrum.

The relationship between wavelength, frequency, and velocity of electromagnetic radiation (including light) is given by the well-known equation

$$f\lambda = c \qquad (2.1)$$

where f is the frequency (in Hz = s^{-1}), λ is the wavelength (in m), and c is the velocity of light (in m s^{-1}). Inherent in Maxwell's equations of electromagnetism is the expression from which the velocity of light in a vacuum can be calculated. This is given by

$$c = \frac{1}{\sqrt{\varepsilon_0 \mu_0}}, \qquad (2.2)$$

where $\varepsilon_0 = 8.85 \times 10^{-12}$ C^2 N^{-1} m^{-2} is the permittivity in a vacuum, and $\mu_0 = 4\pi \times 10^{-7}$ T· m A^{-1} is the permeability in a vacuum, both given in the International System (SI) of units (C for coulomb, N for newton). Substituting the values for ε_0 and μ_0 into Eq. (2.2) gives the result $c = 3.00 \times 10^8$ m s^{-1} = 186,364 mi s^{-1} as an accurate measure of the speed of light in matter-free space (vacuum).

6 RELATIVISTIC FLIGHT MECHANICS AND SPACE TRAVEL

In the middle of the nineteenth century, two French physicists, Armond Fizeau and Jean Foucault, measured the velocity of light in air. From their results, the velocity of light was found to be rather precisely 3×10^8 m s^{-1}. This, of course, agrees with the theoretical value in a vacuum obtained from the application of Eq. (2.2). More importantly to the goals of this book, Fizeau measured the velocity of light in stationary and moving liquid media. He found that the velocity of light in water was reduced to about 3/4 of its velocity in air. In fact, he found that the velocity of light, c, in any liquid at rest was equal to c/n where n is the refractive index of the liquid (equal to unity in a vacuum). Furthermore, Fizeau also found that the velocity of light was influenced by a high flow velocity of the liquid (about 30 m s^{-1}) depending on whether the liquid was moving with or against the propagation direction of light. Phenomenologically, this was viewed as a "drag" on the velocity of light which either increased or decreased a little from the stationary liquid value of c/n depending on the liquid flow with or against the propagation direction of light, respectively. These experiments provided the backdrop for the famed Michelson–Morley experiment that follows.

In 1887, about 36 years following Fizeau's experiments, the physicist Albert Michelson and his assistant Edward Morley set about to show that the velocity of light would change if that light were to propagate through an ether wind thought to exist by virtue of the earth's motion about the sun. The thinking of the time was that the influence of the ether wind (see Glossary) would affect the velocity of light in a manner similar to the effect on the light propagation through a moving liquid as demonstrated by Fizeau. Hence, if no ether wind existed, no velocity change would be observed.

To better understand the Michelson–Morley experiment, we consider a thought experiment featuring two round trips of a boat in a river. First, let the round trip be between two points A' and B' parallel to the bank of the river, then let the second round trip be between two points on opposite sides of the river. These two round trips are depicted in Figs. 2.1(a) and (b), respectively, where v is the surface velocity of the river, V is the velocity of the boat, and L is the distance between the two points A' and B' and between the two points A and B. In Fig. 2.1(a) the time for the round trip is easily found to be

$$t_{A' \leftrightarrow B'} = \frac{L}{V+v} + \frac{L}{V-v} = \frac{2L/V}{(1-v^2/V^2)} \quad (2.3)$$

for $v < V$.

Now, if the boat is launched at point A to reach point B in Fig. 2.1(b), it must move along a path to point C so as to compensate for river current, v, when the boat velocity is again set at V. Noting that distances and velocities form similar right triangles, it follows that $BC/AC = v/V$ so that

$$(AC)^2 = (AB)^2 + (BC)^2 = (AB)^2 + \left[(AC)^2 \cdot \frac{v^2}{V^2}\right].$$

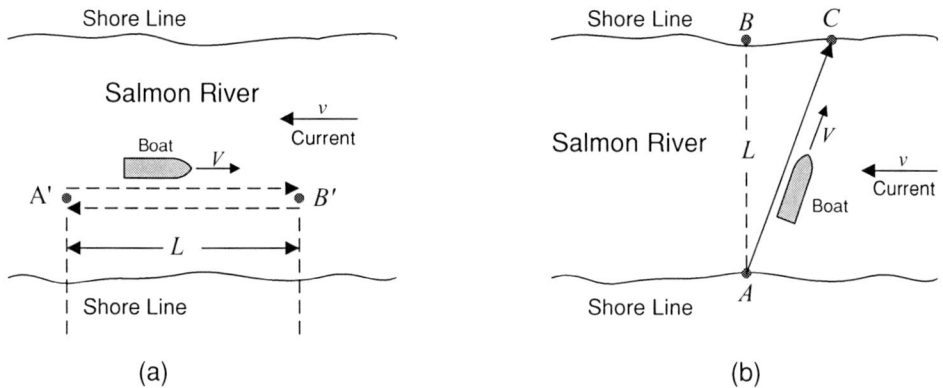

FIGURE 2.1: (a) Boat traveling with velocity $V - v$ against the current and $V + v$ with the current for a round trip between A' and B'. (b) Boat launched at A so as to reach point B must move along a path to point C in order to compensate for the river flow rate, v. A similar compensatory path is necessary for boat launched at B to reach point A

Solving for $(AC)^2$ gives the result

$$(AC)^2 = \frac{(AB)^2}{1 - v^2/V^2}$$

or

$$(AC) = \frac{(AB)}{\sqrt{1 - v^2/V^2}} = \frac{L}{\sqrt{1 - v^2/V^2}}. \quad (2.4)$$

Thus, we see that the distance (AC) is larger than (AB) by a factor $1/\sqrt{1 - v^2/V^2}$ for $v < V$ (remember that the factor $\sqrt{1 - v^2/V^2} < 1$ always). Then for a round trip, the time expended would be $2(AC)/V$ giving

$$t_{A \updownarrow B} = \frac{2L/V}{\sqrt{1 - v^2/V^2}}. \quad (2.5)$$

This means that $t_{A' \leftrightarrow B'}$ takes longer than $t_{A \updownarrow B}$ by the factor $(1 - v^2/V^2)^{-1/2}$, and that both these time durations take longer than that required for a boat round trip across still water $(2L/V)$.

2.2 MICHELSON–MORLEY EXPERIMENT AND THE NULL RESULT

The simplified Michelson–Morley experimental setup presented in Fig. 2.2 is not far removed, conceptually speaking, from the boat/river analogy just discussed. Here, a light beam is issued at the light source to the half-silvered mirror, M_1. At M_1 the light beam is partially reflected

8 RELATIVISTIC FLIGHT MECHANICS AND SPACE TRAVEL

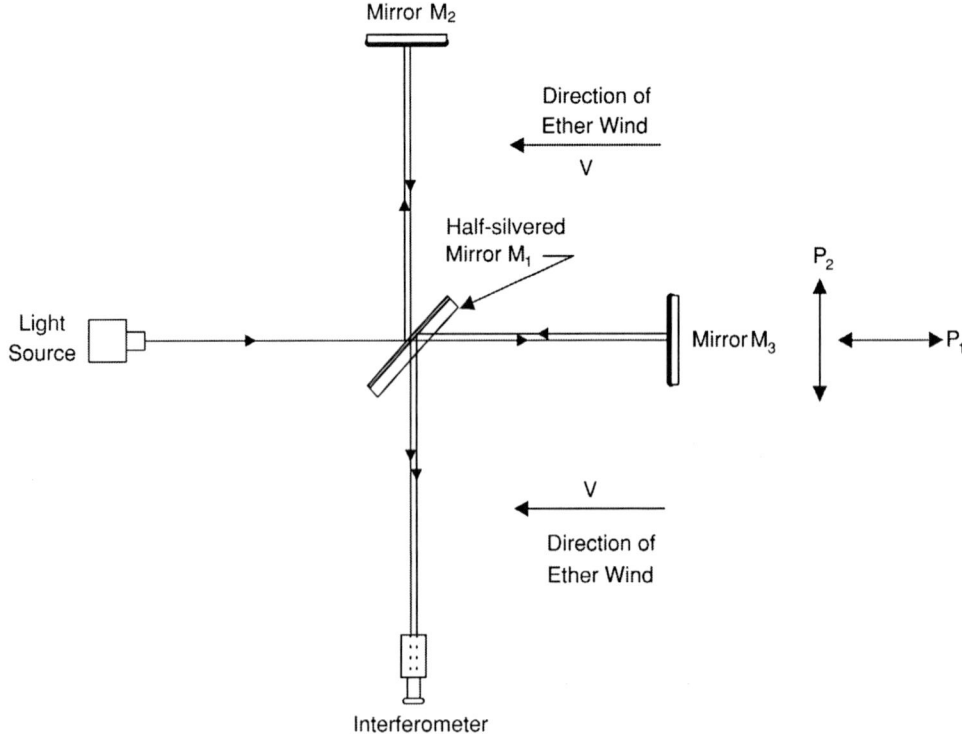

FIGURE 2.2: Simplified Michelson–Morley experimental setup showing two paths of light, path P_1 and path P_2. Path P_1 is against and with a possible ether wind, and path P_2 is transverse to it

to mirror M_2 and partially transmitted to mirror M_3 whereupon the beams are reflected back from these mirrors to M_1. At M_1 the reflected beam from M_2 is partially transmitted to the interferometer while the reflected beam from M_3 is partially reflected to the same interferometer. All light paths being quite long and of equal length, the two beams entering the interferometer should be in constructive interference with maximum illumination if no ether wind were present. On the other hand, the presence of an ether wind would produce a partial destructive interference due to a small relative delay in the light path P_2 transverse to the ether wind moving with a velocity, v.

Returning to the boat/river analogy, the estimated time ratio for light traveling transverse to that traveling longitudinally against and with the ether wind would be

$$\frac{t_\updownarrow}{t_\leftrightarrow} = \frac{2L/c\sqrt{1-v^2/c^2}}{2L/c(1-v^2/c^2)} = \frac{1-v^2/c^2}{\sqrt{1-v^2/c^2}} = \sqrt{1-v^2/c^2} \qquad (2.6)$$

where a simple substitution of c for V has been made. This means that the light propagation time measured transverse to the ether wind would take less time by a factor $\sqrt{1-v^2/c^2}$ than the light propagation time measured against and with the ether wind. Given that the speed of the earth in orbit around the sun is approximately 3×10^4 m s^{-1} (6.7×10^4 mi h^{-1}), and that this can be taken as the speed of the ether wind, v, then the time contraction factor in Eq. (2.6) becomes approximately $\sqrt{1 - (3 \times 10^4)^2/(3 \times 10^8)^2} = \sqrt{1 - 10^{-8}}$, obviously much too small to have been measured directly. However, by use of the Michelson interferometer in the MM experimental setup of Fig. 2.2, interference fringe shifts could be measured with more than sufficient accuracy to detect an ether wind effect. Thus, if an ether wind were present, there would be a resulting partial destructive interference detectable by the interferometer.

The results of the Michelson–Morley experiment (published in 1887) showed *no* partial destructive interference—the NULL result. This experiment has been corroborated numerous times by experiments performed on and above the earth's surface all with the same null result thereby leaving no doubt as to its veracity. Thus, the conclusion is as follows:

No ether wind exists and the velocity of light is, in fact, constant and seemingly independent of motion of the light source.

This conclusion does not mean that the null result of the Michelson–Morley experiment was not without alternative explanations. In fact, it was a British physicist G. F. Fitzgerald that proposed in the 1890s that all bodies in motion, including measuring instruments such as interferometers, underwent a contraction (shrinkage) by a factor $\sqrt{1 - v^2/c^2}$. He used this proposal to reconcile the presence of the ether wind with the null result of the Michelson–Morley experiment. This led to the following nonsense limerick of unknown authorship that satirizes length contraction theory:

There was a young fellow named Fiske
Whose fencing was strikingly brisk;
So fast was his action
The Fitzgerald contraction
Reduced his rapier to a disk.

Actually, the contraction factor $\sqrt{1 - v^2/c^2}$ was also proposed independently by Dutch physicist Hendrik Lorentz in the 1890s to explain theoretically the null result of the Michelson–Morley experiment. As it turns out, however, the contraction factor by Fitzgerald and Lorentz was, at best, an incomplete if not incorrect explanation of the ramifications resulting from the Michelson–Morley experiment. It would be left to the revolutionary work of Albert Einstein in 1905 to provide a detailed explanation of the Michelson–Morley experiment—and much more!

2.3 POSTULATES OF EINSTEIN'S SPECIAL THEORY OF RELATIVITY

In 1905, Albert Einstein, at the age of 26 and then an employee of the Swiss Patent Office, published four articles in the German journal *Annalen der Physick* (*Annals of Physics*). The first article was titled "On a Heuristic View Concerning the Production and Transformation of Light." It provided a theoretical explanation of the experimentally observed photoelectric effect, the ejection of electrons from a metal surface due to incident light on the metal. Actually, this first article introduced the concept that light consists of independent packets (particles) of energy Einstein called "light quanta" now referred to as photons. It was this first article—not his work on relativity—for which he received the Nobel Prize in Physics in 1921.

The second published article, titled "On a New Determination of Molecular Dimensions," Einstein had written as his doctoral dissertation. In the third article titled "On the Movement of Small Particles Suspended in Stationary Liquids Required by the Molecular-Kinetic Theory of Heat," Einstein explained theoretically the Brownian motion of pollen particles in liquids due to the thermal motion of atoms observed by British botanist Robert Brown. From such data Einstein was able to estimate the size of atoms and confirm the atomic theory of matter which, at the time, was still a matter of scientific debate.

The fourth paper, with the interesting title "On the Electrodynamics of Moving Bodies," was to shake the scientific world—his first paper on the theory of relativity now known as Einstein's Special Theory of Relativity. He wrote this 9000 words article in just five weeks, a treatise which was to become as comprehensive and revolutionary as Isaac Newton's *Principia*, and to depose Newton's conception of time. In this last paper Einstein dispensed with the idea of the ether, thought to fill all space, and the need for a single absolute reference frame at rest. Furthermore, Einstein's fourth paper revealed the important relationships between time and space, and between energy and mass (expressed by $E = mc^2$). He also established the relativity of velocity, time, and distance, and predicted the composition of relativistic velocities—all implying that only relative motion is important. Inherent in this paper are two postulates that may be summarized as follows:

Postulate I (*Relativity Principle*). The laws of science are valid in all inertial reference frames—that is, all inertial reference frames are equivalent.

Postulate II (*The Constancy of the Speed of Light Principle*). Light propagates through matter-free space at a definite velocity c independent of the state of motion of the emitting source or observer.

An inertial reference frame is defined as one that obeys Newton's first law of motion (the law of inertia), namely a reference frame at rest or one of uniform motion along a straight path. A rocket traveling with a constant velocity along a straight path would be viewed as being an inertial frame of reference. Strictly speaking, the earth, on which we live, is not an inertial reference frame since it is rotating about its axis (creating a pseudocentrifugal force against earth's gravity) and is subjected to fluctuating gravitational forces produced by its moon and sun as it revolves about the sun. However, the forces to which the earth is subjected are so small that we can and will consider the earth as an inertial reference frame. Thus, the laws of physics are valid on earth as they are in any inertial reference frame. In contrast, any spaceship that is accelerating in matter-free and gravity-free space away from or toward the earth will be considered a noninertial frame of reference. In the development of relativistic flight mechanics, advantage will be taken of these assertions—the earth will be considered as an inertial rest frame relative to a space vehicle accelerating at relativistic speeds taken as a noninertial frame of reference.

Postulate II may appear to be illogical according to our everyday experiences. But remember that the null result of the Michelson–Morley experiment requires that there be no ether that fills all of space, hence no ether wind flowing about any moving object in space. Now consider a simple thought experiment where a flashlight, traveling in free space at the speed of light c emits a beam of light in the direction of its path of motion. According to Postulate II, the beam of light emitted from the flashlight is still c, not $2c$. As we move through the development of relativistic flight mechanics, it will be important to keep the validity of Postulate II in mind. The composition of velocities can be determined on the Newtonian model of a single absolute reference frame only if the velocities (to be added) are much less than the velocity of light, c. When the velocities are relativistic, meaning that they are some whole fraction of c, then two reference frames are necessary to determine the composition of velocities. This will be explored later in this chapter.

2.4   PRINCIPLE OF SIMULTANEITY AND SYNCHRONICITY

There now arises the question as to whether or not two events taking place at different positions in space can be taken as simultaneous by an observer positioned somewhere else in space. To understand the problem of detecting simultaneous events, consider a thought experiment common to our earth experiences. Let two people be within sight of each other at fixed points A and B. If person A (at point A) flashes a light signal in the direction of person B (at point B) and that light signal is reflected back to A by a mirror at B, we can say that for all practical purposes those events are simultaneous, even if it is not possible to measure accurately the light propagation times involved. Communication between person A and person B has been at the speed of light and over a short distance. In this case, there is no need to define time or ponder the problem of understanding the concept of "time."

Now, let us move out into space where we again consider two points, one within spaceship A and the other in spaceship B both moving at a uniform velocity v in the same direction, but

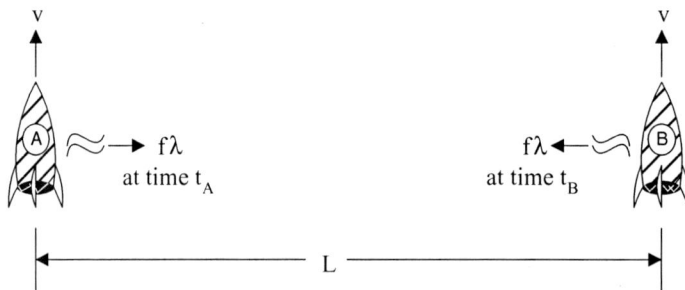

FIGURE 2.3: Test of synchronicity of chronometers (clocks) in space ships A and B having uniform motion v and of constant separation L

widely separated from each other at a constant astronomical distance, L, as shown in Fig. 2.3. Let us assume that a light signal is issued from spaceship A at the absolute time t_A as read by a chronometer (clock) in A. Let this light signal arrive at spaceship B and be immediately reflected back in the direction of A at the absolute time t_B as read by a chronometer in B. Then finally assume that the light signal arrives back at spaceship A at the absolute time t'_A as read by the chronometer in rocket A. Now assert that the time the light signal takes to travel from spaceship A to spaceship B is the same as the time it takes for the signal to travel from spaceship B to spaceship A, hence $t_{AB} = t_{BA}$. Under these assumptions and conditions, the chronometers (clocks) in spaceships A and B are synchronized to each other if, and only if,

$$t_B - t_A = t'_A - t_B = (t_A + 2t_{AB}) - t_B.$$

Again, we have asserted that the transit times $t_{AB} = t_{BA}$ over the fixed distance L. This expression is easily proved by taking $t_A = 0$ giving $t_B = t_{AB}$ as required if clocks A and B are synchronized.

From the above and in agreement with Postulate II, we find that

$$\frac{2L}{t'_A - t_A} = c. \qquad (2.7)$$

Let us pause to reflect on what we have learned. First, it should be evident that the *simultaneity* of events taking place at separate points in space must be defined in terms of the *synchronicity* of the clocks at those points. Having established the synchronicity requirements for the clocks in spaceships A and B of Fig. 2.3, it is now possible to determine whether or not events taking place in these spaceships are simultaneous—a time element common to both spaceships, e.g., transit times $t_{AB} = t_{BA}$, had to be known. In a sense, this is the definition of time.

Take a third point C in space of Fig. 2.3 and declare that a clock at point C is synchronous with the clock at A. Since clocks in A and B are synchronized to each other, it follows that the clock at C is synchronized with that at B. What this implies is that point C is moving at a uniform velocity v along with and in the same direction as spaceships A and B. Thus, any number of points in a given inertial reference frame in space can be synchronous with each other if a time element is common between all of them. Conversely, clocks synchronized in system S_1 will not appear to be synchronized when observed from another system S_2 in motion with respect to S_1.

Consider again Fig. 2.3, and take the case where spaceship A accelerates, eventually reaching instantaneous velocities approaching the speed of light, c, and let spaceship B remain at uniform velocity relative to spaceship A. Spaceship A must now rely on measurements taken independently of spaceship B since it is a noninertial frame of reference. Though their clocks may have been synchronized prior to spaceship A's departure from B, as in Fig. 2.3, those clocks are no longer synchronized and simultaneous events cannot be measured. This matter will be discussed in some detail in Chapter 4 where Earth takes the place of spaceship B.

2.5 RELATIVISTIC TIME DILATION AND THE CLOCK PARADOX

Shown in Fig. 2.4 is another thought experiment; one that is commonly used to illustrate relativistic *time dilation*. In Fig. 2.4(a), a spaceship moves at constant high velocity v along a straight path. At some point in time the spaceship emits a light signal which is reflected by a fictitious mirror located somewhere in space at a constant distance L from the light source. (Note that the mirror is considered part of the spaceships reference frame and, therefore, could have been mounted within the spaceship at a distance L from the light source.) The reflected beam is captured by a light sensor located next to the light source and the time is recorded by a chronometer (clock). The time elapsing between emission and sensing of the light beam is called transit time and is calculated by the astronauts to be

$$\Delta t_e = \frac{2L}{c}. \qquad (2.8)$$

The calculated value for Δt_e in Eq. (2.8) will be found to be in excellent agreement with that measured with the chronometer by the astronauts in the spacecraft. In this book, we will use subscript e to indicate an *eigenvalue* or *characteristic (proper value)* of any quantity that is measured or calculated by astronauts in spaceships traveling at high velocities relative to a rest frame of reference. In this way, proper values can be distinguished from those measured or calculated by an observer in the rest frame of reference. Thus, the symbol Δt_e is the proper transit time of the light beam as calculated or measured by the astronauts.

In effect, the astronauts in Fig. 2.4(a) need not be aware that their spaceship is moving since their measurements are made within their own *inertial* frame of reference, the spaceship.

14 RELATIVISTIC FLIGHT MECHANICS AND SPACE TRAVEL

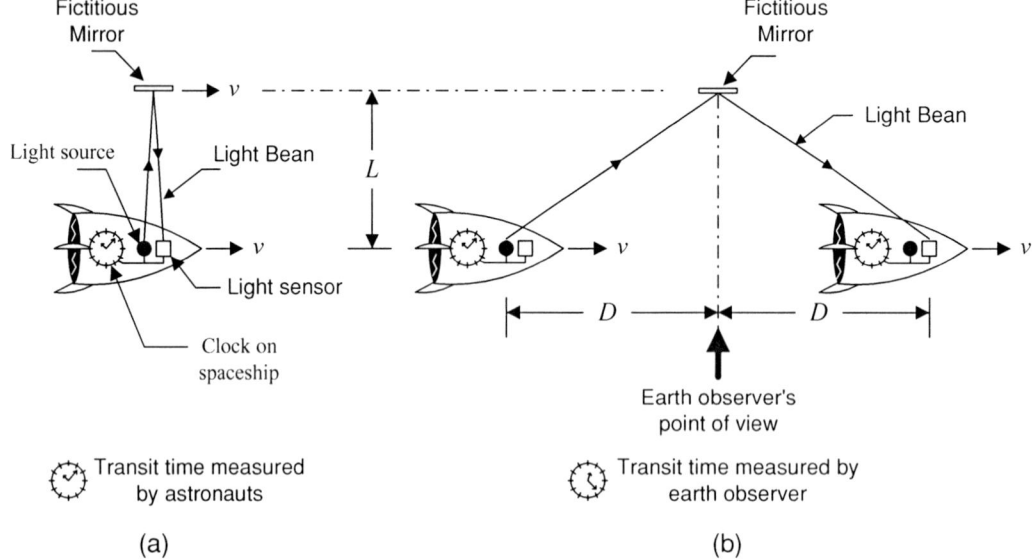

FIGURE 2.4: (a) Transit time of light beam as measured by astronauts. (b) Transit time of light beam as measured by an earth observer

From the point of view of the earth observer, this is no longer true. In this case, since the spaceship has traveled a distance $2D$, the earth observer measures the time elapsing between light emission and light sensing in Fig. 2.4(b) to be

$$\Delta t = \frac{2\sqrt{L^2 + D^2}}{c}. \qquad (2.9)$$

But the time it takes for the astronauts to travel the distance $2D$ at a velocity v, is evidently

$$\Delta t = \frac{2D}{v}. \qquad (2.10)$$

Introducing Eqs. (2.8) and (2.10) into Eq. (2.9) gives the results

$$\Delta t = \frac{2\sqrt{(c^2 \Delta t_e^2/4) + (v^2 \Delta t^2/4)}}{c}$$

which simplifies to

$$\Delta t^2 = \frac{c^2 \Delta t_e^2 + v^2 \Delta t^2}{c^2} = \Delta t_e^2 + \frac{v^2}{c^2}\Delta t^2$$

giving $\Delta t = \dfrac{\Delta t_e}{\sqrt{1 - v^2/c^2}}$ or in differential form $dt = \dfrac{dt_e}{\sqrt{1 - v^2/c^2}}. \qquad (2.11)$

This important result of Einstein's special theory of relativity is commonly known as *time retardation* or *time dilation—the slowing of moving clocks*. It states simply that clocks moving at relativistic velocities run more slowly than do clocks that are at rest—time dilates. Thus, at relativistic velocities (i.e., as v approaches c) $\Delta t \gg \Delta t_e$, since $\sqrt{1 - v^2/c^2}$ approaches zero. Clearly, Eq. (2.11) requires that $v < c$ since if $v = c$ then $\Delta t = \infty$ or $\Delta t_e = 0$. Or if $v > c$, in violation of Postulate II, the denominator becomes an imaginary number, $\sqrt{-x}$. Also, if $v \ll c$, then $\Delta t \to \Delta t_e$ and the calculations resolve into a Galilean–Newtonian model. In short, Eq. (2.11) is applicable to objects having rest mass (e.g., a subatomic particles, spacecrafts, etc.) that move at relativistic velocities. But time dilation is not restricted to the slowing of clocks in spaceships or to the increasing of half-lives of subatomic particles. Time dilation also applies to all of nature's clocks, including human biological clocks, causing them to slow down in spaceships moving at relativistic velocities relative to some reference frame at rest. This, of course, leads to the subject of the *clock paradox* or *twins paradox* considered next.

The clock or twins paradox is probably not a paradox at all. What you might already have guessed from the above discussion is that all clocks, especially biological clocks, run slower on a spacecraft moving at relativistic velocities when compared to those say on earth from which the spacecraft originated. What does this mean? It simply means that the astronauts in spaceships traveling at relativistic velocities have rhythmic heartbeats pulsing slower than those of their stay-at-home earth observers. Yes, they will check their vital signs only to find that the data appear to be in agreement with those measured before departure from the earth. Now, let us consider the case of identical twins, one an astronaut and the other a scientist. Suppose that a space flight is planned and the astronaut will be onboard the spacecraft while his scientist twin will stay on earth. Before departure, the twins will precisely synchronize their Rolex watches and chronometers to each other. The spacecraft launches and accelerates to attain relativistic velocities on its path to some point in space whereupon it decelerates and eventually reverses direction. The spacecraft then accelerates to achieve relativistic velocities and returns to earth with the appropriate deceleration to achieve a safe landing. On return, the twins compare dates and times from their chronometers or watches only to find that the say-at-home twin aged more than his astronaut twin. The returning astronaut will show an elapsed time interval

$$\Delta t_e = \int_{t_1}^{t_2} \left(1 - v^2(t)/c^2\right)^{1/2} dt$$

which would become $\Delta t_e = \Delta t \sqrt{1 - v^2/c^2}$ if v were constant assuming that the acceleration and deceleration periods were small compared to the period of constant relativistic velocity v, i.e., if $v \neq v(t)$. In comparison, the stay-at-home twin would record an elapsed time of $t_2 - t_1 > \Delta t_e$.

16 RELATIVISTIC FLIGHT MECHANICS AND SPACE TRAVEL

TABLE 2.1: Representative Values of the Dilation and Contraction Factors as a Function of v/c

v/c	0.2	0.3	0.4	0.5	0.6	0.8	0.9	0.99	0.999	0.9999	0.99999
$(1 - v^2/c^2)^{-1/2}$	1.02	1.05	1.09	1.16	1.25	1.67	2.29	7.09	22.4	70.7	223.6
$(1 - v^2/c^2)^{1/2}$	0.98	0.95	0.92	0.87	0.80	0.60	0.44	0.14	0.045	0.014	0.0045

Shown in Fig. 2.5 are plots of the dilation and contraction factors as functions of v/c. In the former case of the dilation factor $(1 - v^2/c^2)^{-1/2}$, little change from unity takes place until v/c exceeds about 0.4 and then reaches a significant value only after v/c exceeds 0.8. Representative values for the dilation and contraction factors are given in Table 2.1 as a function of v/c. As will be apparent in the following discussions, these relativistic factors are not only important in the discussions of relativistic time and distance but also enter the treatment of mass, momentum, and energy. For this reason, it will be useful to keep in mind the information contained in Fig. 2.5 and in Table 2.1.

Time dilation, on which the twins paradox is based, has been confirmed by numerous experiments. Among these was an experiment performed in 1971. In this experiment, extremely accurate atomic clocks were flown around the earth in jet planes. The atomic clocks, having an accuracy in the nanosecond (10^{-9} s) range, confirmed the validity of Eq. (2.11). Decades earlier, experiments on cosmic ray μ-mesons (muons) entering the earth's lower atmosphere at nearly

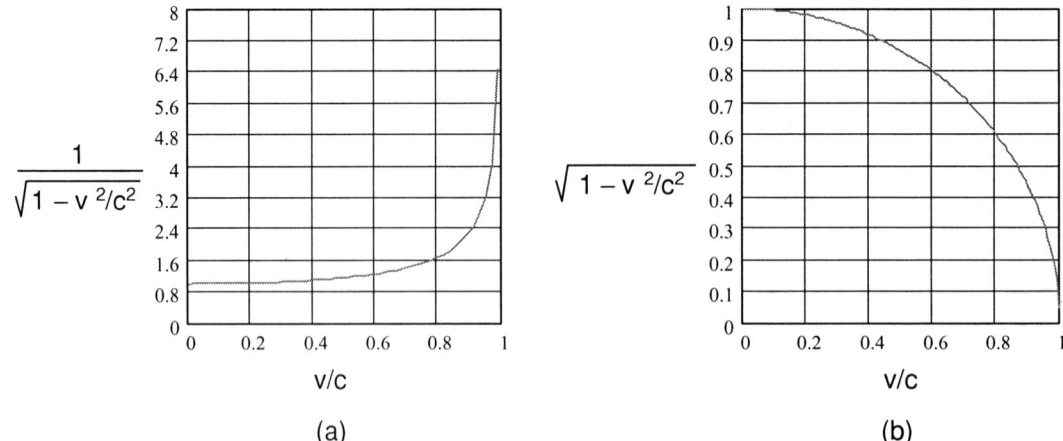

FIGURE 2.5: (a) Relativistic time dilation factor as a function of v/c showing little change in the dilation factor from 0 up to about $v/c = 0.4$ and significant change beyond $v/c = 0.8$. (b) Relativistic contraction factor as a function of v/c

the speed of light experienced increased life-times over those muons at rest, all in agreement with time dilation expressed by Eq. (2.11).

Like distance contraction (see the end of Section 2.2), relativistic time dilation has had its influence on the literature and is perhaps poorly satirized by the following nonsense limerick of unknown authorship:

There once was a lady called Bright
Who could travel faster than light;
She went out one day
In a relative way
And came back the previous night.

Clearly, this limerick is really not a valid characterization of either time dilation or contraction.

2.6 RELATIVISTIC DISTANCE CONTRACTION

Before Einstein, the accepted thinking was that time and space were completely separate entities. This was made clear in Newton's *Principia*, the leading reference for scientific thought of the time. This was dramatically changed by publication of Einstein's paper "On the Electrodynamics of Moving Bodies." Here, Einstein brought forth the important notions that space and time are inseparable, and that there is no universal quantity called "time" that all clocks would measure. This leads immediately to the following realization: if time dilates for clocks moving at relativistic velocities, then distance must contract under the same conditions. This, of course, is what Fitzgerald and Lorentz proposed in the 1890s but for different reasons.

To better understand the concept of distance contraction, let us engage in another simple thought experiment. Consider a spacecraft, launched from the earth and accelerated to relativistic velocity v on a path to star X, as illustrated in Fig. 2.6; note the difference in how the earth observer and astronaut view the spacecraft's motion, as depicted in Figs. 2.6(a) and (b). Thus, it makes no difference whether the spaceship is viewed as being in motion with respect to earth and star (as viewed from the earth) or whether the spaceship is stationary in space while the star is approaching and the earth is receding (as viewed by the astronauts). But it is the spaceship, not the earth that is traveling at relativistic velocities. To continue, assume that the acceleration travel distance to achieve a constant relativistic velocity v is negligible compared to the total distance S (measured from the earth) of the star from the earth. From the point of view of an earth observer in Fig. 2.6(a), the elemental time required for the spacecraft to travel a distance dS on a trajectory to the distant star X would be

$$dt = dS/v. \qquad (2.12)$$

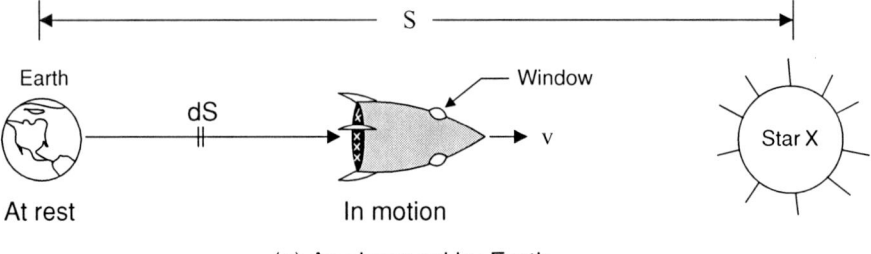

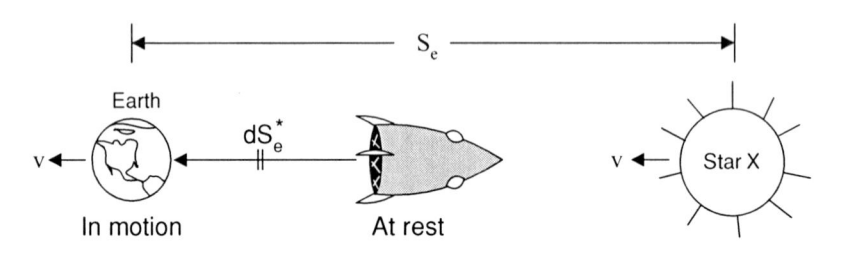

18 RELATIVISTIC FLIGHT MECHANICS AND SPACE TRAVEL

FIGURE 2.6: Two points of view of the spacecraft's journey from Earth to star X. (a) View of the spacecraft as observed from Earth. (b) View of earth by the astronauts showing star advancing toward the spacecraft at a velocity v, and earth receding from the spacecraft at the same velocity

From the point of view of the astronauts in the spacecraft of Fig. 2.6(b) traveling at relativistic velocity v, the time element dt_e required for the earth to move an elemental distance dS_e^* (away from the spacecraft as in Fig. 2.6(b)) would be less than the time observed on earth for the spacecraft to move an elemental distance dS. Thus, by the relativistic time contraction we have

$$dt_e = dt\sqrt{1 - v^2/c^2} = dS_e^*/v. \qquad (2.13)$$

Introducing Eq. (2.12) into Eq. (2.13) gives

$$(dS/v)\sqrt{1 - v^2/c^2} = dS_e^*/v$$

or

$$dS_e^* = dS\sqrt{1 - v^2/c^2}. \qquad (2.14)$$

Equation (2.14) shows that the elemental earth's distance along the direction of earth's motion, dS_e^*, as determined by the astronauts, is shorter than dS observed from the earth to the spacecraft by a factor of $\sqrt{1 - v^2/c^2}$, which is significant only as v approaches c—*relativistic*

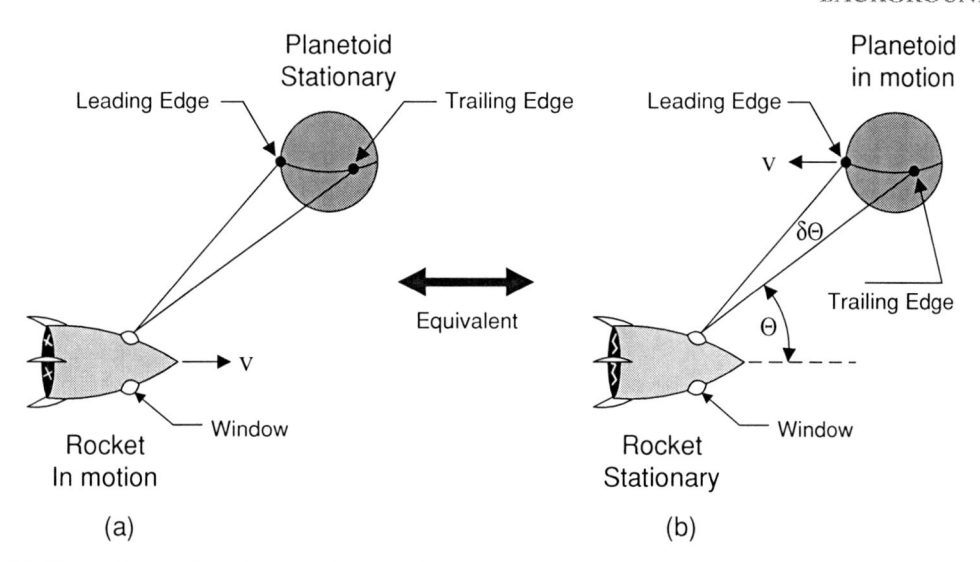

FIGURE 2.7: View of a spherical planetoid by astronauts in a rocket passing by at relativistic velocity, v. (a) Rocket in motion relative to stationary planetoid. (b) Equivalently, planetoid in motion relative to stationary rocket

distance contraction. At nonrelativistic velocities the distances would be the same, $dS^* = dS$. Here again, the subscript e is used to indicate an *eigen* (or *characteristic*) quantity, a quantity that is characteristic of what is observed by astronauts in a spacecraft. The word *proper* will be used interchangeably with *characteristic* to indicate an *eigen* quantity. The total proper distance to the star observed by the astronauts, S_e, is less than the total distance observed by the earth, S, and will be discussed later in Section 4.4.

At this point we must deal with a common misconception with regard to the application of Eq. (2.14). Consider the spaceship in Fig. 2.7(a) moving at relativistic velocity, say $v = 0.95c$, in passing near a spherical planetoid. Note that it makes no difference whether the spaceship is moving and the planetoid is stationary or vice versa; relativistic time and distance are symmetrical effects. Thus, Fig. 2.7(b) is the equivalent of Fig. 2.7(a). Let the astronauts be stationed at the windows so as to view their surrounding progress with instruments available to them in the spaceship. The question is how will the planetoid appear to the astronauts as it approaches and passes by the planetoid? Will it appear to be distorted because of Eq. (2.14)? One might be tempted to believe that this body would appear vertically oblate, i.e., an ellipsoidal form having its minor axis (along the direction of trajectory) less than its unchanged major (vertical) axis as a result of Eq. (2.14). However, because of the finite velocity of light and time dilation, the light from the leading and trailing edges of the sphere reach the astronaut's window at slightly different times thereby canceling the effect of relativistic distance contraction. So the spherical body retains its perfectly spherical appearance to the astronauts as the spaceship passes nearby at relativistic velocity $v = 0.95c$.

To see how this is possible, we will engage in another simple thought experiment. Let δt be the difference in the time a light pulse takes to propagate to the astronaut's window from the leading versus the trailing edge of the planetoid as measured along its direction of motion shown in Fig. 2.7(b). But during this time $\delta t = \delta t_0/\sqrt{1 - v^2/c^2}$, subject to time dilation given by Eq. (2.11), the planetoid has traveled a distance $\delta S = v(\delta t)\sqrt{1 - v^2/c^2}$ subject to distance contraction given by Eq. (2.14). Combining these equations gives the simple result

$$\delta S = v(\delta t)\sqrt{1 - v^2/c^2}$$
$$= v(\delta t_0)\frac{\sqrt{1 - v^2/c^2}}{\sqrt{1 - v^2/c^2}} = v(\delta t_0).$$

Here $v(\delta t_0)$ is the same result as if both the spacecraft and planetoid were fixed within the same inertial reference frame with respect to each other. Hence, the planetoid retains its spherical appearance unaffected by distance contraction effects. Similar results would be obtained if the planetoid were replaced by a long, slowly rotating flat ruler. The ruler would appear unchanged in any of its orientations as viewed by the astronauts. Actually, distance contraction effects could be observed by the astronauts were they to photograph the passage of the planetoid or ruler by means of a camera with a lens and shutter larger than the size (in the direction of motion) of the object being photographed. So for long distance (near point) viewing, as elemental angle $\delta\theta$ approaches zero, the minor axis length of the planetoid would be contracted along the direction of motion at relativistic velocities according to $\ell = \ell_0\sqrt{1 - (v^2/c^2)\cos\theta}$. Thus, a completed photograph would reveal this shortened dimension along the direction of trajectory, except at right angle viewing ($\theta = 90°$) for which $\ell = \ell_0$. Historically, it is interesting to note that this simple explanation eluded physicists for more than 50 years following Einstein's historical 1905 paper "On the Electrodynamics of Moving Bodies," now known as the Special Theory of Relativity.

2.7 RELATIVISTIC TRANSFORMATION OF COORDINATES AND ADDITION OF VELOCITIES

The Lorentz transformation equations. We now know that time and space are inseparable, hence *spacetime*. There is no universal quantity called time to which all clocks can be synchronized. Therefore, in order to understand the relativistic transformation of coordinates and the composition of velocities that derive from them, it is necessary to begin with the classical Galilean (after Galileo Galilei) transformation equations and include time as a fourth dimension. Begin by considering the two reference frames illustrated in Fig. 2.8.

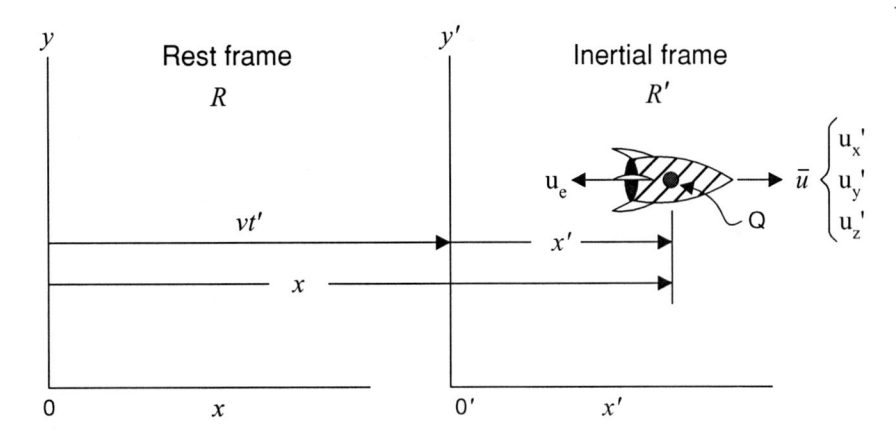

FIGURE 2.8: Two reference frames: one at rest, and the other an inertial frame moving at uniform velocity away from the rest frame along a straight path

In Fig. 2.8, let R represent the rest frame of reference and R' the inertial frame moving at a uniform velocity v away from R along the x-axis. At a time t' the inertial frame R' has moved a distance vt' from R. At $t' = 0$ the two frames of reference coincide so that $0 = 0'$. Within the inertial frame, a spaceship, represented by point Q, moves with a velocity vector $\bar{u}$ in the x' direction with velocity vector components u'_x, u'_y, u'_z. In a time t' the spacecraft has moved a distance $x' + vt'$ from rest frame R. The proper efflux (propellant) velocity is designated as u_e. We will have cause to use the efflux velocity in calculations later in this section.

Knowing the coordinates in inertial frame R', we can write the classical *Galilean transformation equations* for frame R as

$$x = x' + vt'$$
$$y = y'$$
$$z = z'$$
$$t = t' \qquad (2.15)$$

where it is seen that y, z, and t all remain unchanged in a classical system for which motion is only along the x-axis.

Next, suppose point Q in Fig. 2.8 represents a spacecraft traveling with velocity vector components u'_x, u'_y, u'_z in R'. Transforming these velocities in R' to u_x, u_y, u_z relative to rest frame R gives the results

$$u_x = \frac{dx}{dt'} = \frac{d(x' + vt')}{dt'} = u'_x + v$$
$$u_y = u'_y$$
$$u_z = u'_z \qquad (2.16)$$

where again we see that the y and z components remain unchanged in a classical system where motion is only along the x-axis.

Now, let us assume that $u'_x = c$ in Eqs. (2.16); that is, we let the spacecraft travel at the speed of light in the x' direction. Doing this requires that $u_x = c + v$ and that $u_x > c$ for finite v. This violates the basic principle of special relativity that nothing can travel faster than the speed of light (see Postulate II in Section 2.3).

We will need to alter the Galilean transformation equations (2.16) so as to prevent $u_x > c$. To do this requires that we use the following *relativistic transformation equations*,

$$x = \gamma \cdot (x' + vt')$$
$$y = y'$$
$$z = z'$$
$$t = \gamma \cdot (t' + \alpha \cdot v) \tag{2.17}$$

where γ and α are coefficients yet to be determined such that $\gamma \geq 1$.

Before we set about finding γ and α, we will write the inverse of Eqs. (2.17) to give

$$x' = \gamma(x - vt)$$
$$y' = y$$
$$z' = z$$
$$t' = \gamma(t - \alpha \cdot v). \tag{2.18}$$

Here we note that v has been replaced by $-v$, since rest frame R moving to the left of frame R' is equivalent to R' moving to the right of R.

We will now engage in another thought experiment. Let $t = t' = 0$ so as to bring the two coordinate systems of Fig. 2.8 into coincidence such that origins $0 = 0'$. Now allow a light pulse to leave the common origins in R and R' at the same time t. Since c is constant relative to all frames of reference, we can write for x and x' in Eqs. (2.17) and (2.18), respectively, the following two equations:

$$ct = \gamma(ct' + vt') = \gamma(c + v)t' \quad \text{w/r } R$$
$$ct' = \gamma(ct - vt) = \gamma(c - v)t \quad \text{w/r } R'. \tag{2.19}$$

We can solve Eqs. (2.19) simultaneously by introducing $t' = \frac{\gamma}{c}(c - v)t$ from the second equation into the first equation to give

$$ct = \gamma(c + v) \cdot \frac{\gamma}{c}(c - v)t = \frac{\gamma^2}{c}(c^2 - v^2)t.$$

Clearing fractions, canceling t on both sides, dividing by c^2, and taking the square root gives the result

$$\gamma = \frac{1}{\sqrt{1 - v^2/c^2}}. \qquad (2.20)$$

To get α, we introduce x into x', using Eqs. (2.17) and (2.18), with the result

$$x' = \gamma[\gamma(x' + vt') - vt].$$

Then solving for t we have

$$t = \frac{\gamma x'}{v} + \gamma t' - \frac{x'}{\gamma v} \cdot \frac{\gamma}{\gamma} = \gamma\left(t' + \frac{x'}{v} - \frac{x'}{v\gamma^2}\right)$$

$$= \gamma\left[t' + \frac{x'}{v}\left(1 - \frac{1}{\gamma^2}\right)\right]$$

$$= \gamma\left(t' + \frac{x'v}{c^2}\right).$$

This gives the result

$$\alpha = \frac{x'}{c^2} \qquad (2.21)$$

where $(1 - 1/\gamma^2) = v^2/c^2$, from Eq. (2.20), has been introduced into the expression for t to obtain Eq. (2.21).

With these results in mind, we rewrite the transformation Eqs. (2.17) as follows:

$$x = \frac{(x' + vt')}{\sqrt{1 - v^2/c^2}} = \gamma(x' + vt')$$

$$y = y'$$

$$z = z'$$

$$t = \frac{(t' + x'v/c^2)}{\sqrt{1 - v^2/c^2}} = \gamma(t' + x'v/c^2). \qquad (2.22)$$

These relativistic transformation equations have become known as the *Lorentz transformation equations.*

The composition of velocities. We know that u_x cannot be greater than c. So let us test whether this is true by using the relativistic transformation equation for x given in Eqs. (2.22) to obtain the composition of velocities as follows:

$$u_x = \frac{dx}{dt} = \frac{d}{dt}[\gamma(x' + vt')$$

$$= \frac{d}{dt'}[\gamma(x' + vt')]\frac{dt'}{dt} = \gamma\left[\frac{dx'}{dt'} + v\right]\frac{dt'}{dt}.$$

But $dx'/dt' = u'_x$, and $dt'/dt = (dt/dt')^{-1} = [\gamma(1 + \frac{v}{c^2}\frac{dx'}{dt'})]^{-1} = [\gamma(1 + u'_x v/c^2)]^{-1}$. Therefore, the composition of u_x becomes

$$u_x = \frac{\gamma(u'_x + v)}{\gamma(1 + u'_x v/c^2)} = \frac{u'_x + v}{1 + u'_x v/c^2} \quad (2.23)$$

which gives the relativistic sum of u'_x and v relative to rest frame R, where u'_x is the x' component velocity of point Q, and v is the velocity of frame R' relative to R as seen in Fig. 2.8. Now if $u'_x = c$, then $u_x = c$, hence never to allow $u_x > c$. Actually, if either velocity (or both) is c, the result is also $u_x = c$.

Noting that relativistic velocities u_y and u_z are also affected by u_x, they can be obtained in the same manner. From Eqs. (2.22) we can write

$$u_y = \frac{dy}{dt} = \frac{dy'}{dt} = \frac{d}{dt'}(y')\frac{dt'}{dt} = u'_y\left(\frac{dt}{dt'}\right)^{-1} = \frac{u'_y}{\gamma(1 + u'_x v/c^2)}$$

$$u_z = \frac{dz}{dt} = \frac{dz'}{dt} = \frac{d}{dt'}(z')\frac{dt'}{dt} = u'_z\left(\frac{dt}{dt'}\right)^{-1} = \frac{u'_z}{\gamma(1 + u'_x v/c^2)}$$

or collectively,

$$u_x = \frac{u'_x + v}{1 + u'_x v/c^2}$$

$$u_y = \frac{u'_y}{\gamma(1 + u'_x v/c^2)}$$

$$u_z = \frac{u'_z}{\gamma(1 + u'_x v/c^2)}. \quad (2.24)$$

For our purpose, it is necessary to find the composition of spacecraft velocity u'_x along the x' direction. From Fig. 2.8, u'_x is the relativistic sum of velocities v and u_e. Following the form of Eq. (2.23), there results

$$u'_x = \frac{v - u_e}{1 - vu_e/c^2} \quad (2.25)$$

in agreement with the orientation of axes and the fact that u_e is oriented in the opposite direction of v. Thus, the minus sign is the result of two oppositely directed velocity vectors. Here u_e is the proper velocity of the propellant (efflux velocity) as observed by the astronauts in the spacecraft. The fact that u_e exceeds v is of no consequence in the physical sense since, by Newton's second and third laws, the efflux must create a thrust (force) on the spacecraft in the opposite direction, i.e., in the direction of the spacecraft's motion. Equation (2.25) will be used later in Section 3.4.

The relativistic composition of velocities can be generalized for any two velocities as follows:

$$V = \frac{v_1 + v_2}{1 + v_1 v_2/c^2} \quad (2.26)$$

where for low velocities ($v \ll c$) $V = v_1 + v_2$, the classical Newtonian result. Or if either v_1 or v_2 equals c (or both equal c), then $V = c$, but Eq. (2.26) never allows $V > c$.

2.8 RELATIVISTIC MOMENTUM AND MASS

Let us first consider in a thought experiment the classical case of two perfectly elastic balls, ball A and ball B, traveling at nonrelativistic velocities, v_A and v_B, and caused to collide head-on as shown in Fig. 2.9. The conservation of momentum is preserved if $\Delta p_{Before} = \Delta p_{After}$, before and after collision. Therefore, from Fig. 2.9 we write

$$m_A v_A + m_B v_B = m_A v'_A + m_B v'_B.$$

Rearranging gives

$$m_A v'_A - m_A v_A = m_B v'_B - m_B v_B.$$

Solving for m_A there results

$$m_A = m_B \frac{v'_B - v_B}{v'_A - v_A} = m_B \frac{v_B - v'_B}{v_A - v'_A}. \quad (2.27)$$

To continue our thought experiment further, imagine that Fig. 2.9 is altered such that the two balls, ball A and ball B, traveling at different relativistic velocities, are caused to collide and rebound perfectly elastically (at the same velocities) in a glancing fashion, as illustrated in Fig. 2.10. Here, two inertial frames of reference, frame A and frame B, are moving with a velocity v with respect to each other. We assume that the ball masses are a function of their respective velocities and that their rest masses are equal, $m_A = m_B = m_0$. We further assume that Eqs. (2.24) are valid and that $u'_x \to 0$ so the law of conservation of momentum can be

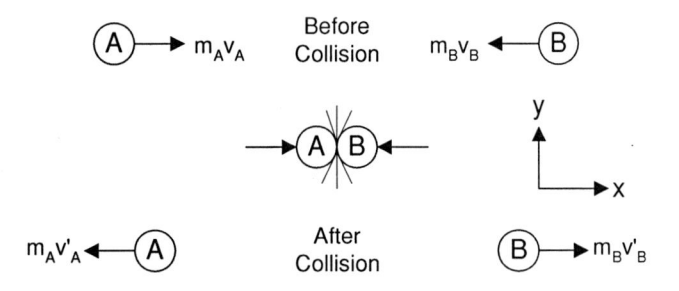

FIGURE 2.9: Collision of two elastic balls to demonstrate the conservation of momentum

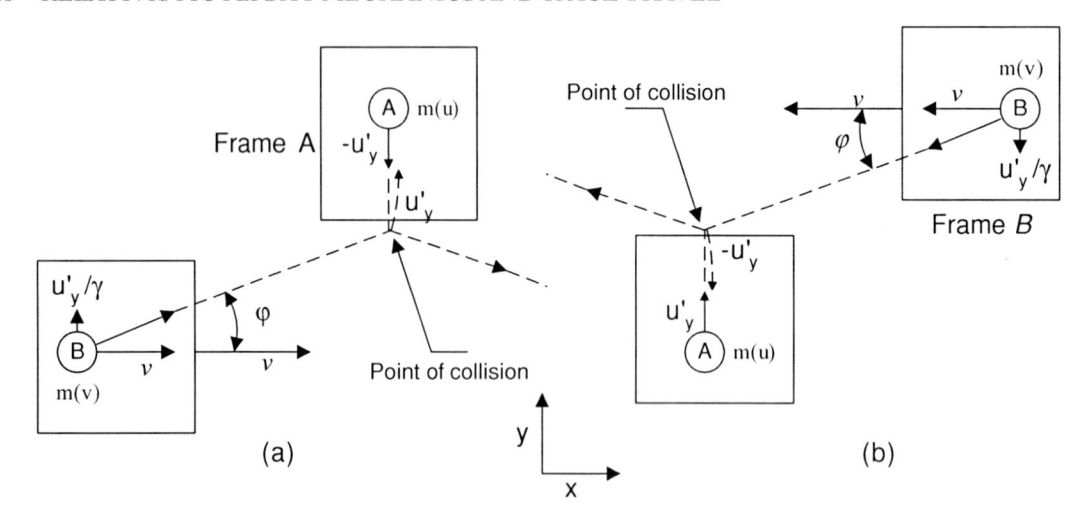

FIGURE 2.10: A perfectly elastic glancing collision of two balls, A and B, from the point of view of two reference frames. (a) As viewed from reference frame A. (b) As viewed from reference frame B

applied along the y-axis. By looking at the collision from the point of view of each frame of reference, we can properly apply the conservation of momentum equating the momentums before with those after the collision event, first with respect to frame A (before) and then frame B (after). For simplicity, let $u'_y = u$ in the equations that follow from Fig. 2.10,

$$-m(u)u + m(v)u/\gamma = -m(v)u/\gamma + m(u)u$$

or solving for $m(v)$ gives

$$m(v) = \gamma m(u) = \frac{m(u)}{\sqrt{1 - v^2/c^2}}. \qquad (2.28)$$

Now let $\varphi \to 0$ so as to cause $u \to 0$ (a perfectly glancing action) with the result that $m(u) \to m_0$, the rest mass. Under this condition Eq. (2.28) becomes

$$m = \frac{m_0}{\sqrt{1 - v^2/c^2}}. \qquad (2.29)$$

Equation (2.29) is valid for an object (subatomic particle or spacecraft) having rest mass m_0 and moving at relativistic velocity v along a straight path independent of any interaction with another object. The perfectly glancing action for which $u'_y = 0$ is equivalent to the motion of the object in a straight trajectory path in matter-free and resistance-free space. Finally, note that as $v \to c$ the mass of the object $m \to \infty$ since $\sqrt{1 - v^2/c^2} \to 0$ for a finite rest mass m_0 of the object.

The relativistic momentum of an object can now be defined as

$$p = mv = \frac{m_0 v}{\sqrt{1 - v^2/c^2}} \tag{2.30}$$

which is normally written in vector form. From Eq. (2.30), we can write *Newton's second law* in its most general form valid at any velocity of the object

$$F = \frac{dp}{dt} = \frac{d}{dt}(mv). \tag{2.31}$$

Clearly, at relativistic velocities the classical form of Newton's second law $\bar{F} = m\bar{a}$ is not valid since $m = m(v)$. As will be recalled, Newton's second law requires that mass, m, be constant. Also, since kinetic energy (E_K) is proportional to dp/dt, it follows from Eqs. (2.30) and (2.31) that as $v \to c$ then $E_K \to \infty$. Thus, to accelerate any object having rest mass to the speed of light would require the expenditure of an infinite amount of energy. We conclude from this that the speed of light is the universal upper limit to which an object of rest mass m_0 can only approach.

It is interesting that Einstein's theory of relativity does not rule out the possibility that $v > c$ for an object having rest mass. Were this true, of course, requires that the contraction factor $\sqrt{1 - v^2/c^2} \to \sqrt{-X}$ be an imaginary quantity. Hypothetical particles with the property $v > c$ have been dubbed "tachyons" after the root word "tachy" meaning "fast." Then, assuming that the rest mass m_0 of such a particle must also be imaginary, it follows that its relativistic mass m must be the ratio of two imaginary numbers and, accordingly, be real. Presumably, the speed of light c would, for such hypothetical particles, become the lower limit of their velocity spectrum.

Of course, no object (particle) has ever been discovered whose velocity is $v > c$. But we bring this subject to the attention of the reader since our knowledge of the universe is still in its infancy. Only in the last couple of decades or so have we been led to believe that the universe consists predominantly of "dark matter" and that "dark energy" appears to be responsible for the accelerating expansion of the visible universe. The point is that there still remains so much about the universe we do not know or understand. What new discoveries await us in the next few decades may be limited only by our collective imagination.

2.9 THE RELATIVISTIC MASS–ENERGY RELATION

If we were to ask someone on the street what they know of Einstein's theory of relativity, they would probably say "well, $E = mc^2$," but with little or no understanding of what it means. In this section, we will develop the mass–energy relations from first principles and conclude with a short discussion of their meaning.

We begin with the generalized work–energy theorem for an object in motion by expressing its kinetic energy in the form

$$E_K = \int_i^f F ds = \int_i^f \frac{dp}{dt} ds = \int_i^f \frac{dp}{dt} \cdot v dt = \int_i^f v \cdot dp \qquad (2.32)$$

since $F = dp/dt$ and $v = ds/dt$, where i and f represent the initial and the final limits of integration, respectively. But $d(pv) = vdp + pdv$ giving the kinetic energy as

$$E_K = \int_i^f d(pv) - \int_0^v p dv \qquad (2.33)$$

where

$$\int_i^f d(pv) = pv]_i^f = mv^2]_0^v = mv^2. \qquad (2.34)$$

Introducing Eqs. (2.34) into Eq. (2.33) and noting that $p = mv$, there results for the kinetic energy

$$E_K = mv^2 - \int_0^v p \cdot dv = mv^2 - \int_0^v \frac{m_0 v}{\sqrt{1 - v^2/c^2}} dv \qquad (2.35)$$

where Eq. (2.30) has been introduced since $m = m(v)$. Now, integrating Eq. (2.35) and introducing Eq. (2.29) gives the results

$$E_K = mv^2 + \left[m_0 c^2 \sqrt{1 - v^2/c^2}\right]_0^v = mv^2 + mc^2(1 - v^2/c^2) - m_0 c^2$$

or finally

$$E_K = mc^2 - m_0 c^2 \qquad (2.36)$$

where $E_0 = m_0 c^2$ is called the *rest energy* of the object. Equation (2.36) shows that the kinetic energy of an object (fundamental particle, spacecraft, etc.) is its total relativistic energy minus its energy at rest. Solving for the total energy of the object results in the well-known *mass–energy expression*

$$E_{Total} = mc^2 = m_0 c^2 + E_K \qquad (2.37)$$

in SI units of Joules, $J = kg\ m^2\ s^{-2}$. Equation (2.37) shows that the total energy consists of the rest energy of the object plus its kinetic energy due to its motion.

A very important consequence of special relativity, as required by Eq. (2.37), is that *mass and energy are equivalent*. It demonstrates that mass is convertible to energy and vice versa—the interconversion of mass and energy. Let us cite some examples. Energy that is produced by fission in nuclear power plants results from mass loss of uranium fuel, and the fusion process taking place on the sun produces radiant energy by the conversion of hydrogen to helium, all according to $E = mc^2$. Neutral pions (π^0) of rest mass m_0^P decay into electromagnetic radiation with the generation of energy exactly equal to $(m_0^P)c^2$ and the disappearance of π^0. Conversely, laboratory experiments have shown that electromagnetic radiation can be converted into electrons.

If matter comes into contact with antimatter, complete mutual annihilation of the matter–antimatter particles occurs with the emission of enormous amounts of electromagnetic energy. Positrons and antiprotons are examples of antimatter that have the same mass as their matter counterparts, electrons and protons respectively, but are of opposite charge. As an example, it is found that when an electron collides with a positron an amount of energy $E_R = 2m_{el}c^2$ is released in the form of electromagnetic radiation accompanied by the complete mutual annihilation of both particles. This demonstrates that all of the rest energy of matter (or antimatter) is potentially transformable into other forms of energy such as gamma radiation or heat. In fact, the interconversion concept applies to matter–antimatter particles as well. If a gamma-ray collides with and is absorbed by an atomic nucleus, its energy is transformed into the rest energies of electron–positron pairs which are created in the process.

So what really is this rest energy m_0c^2? At absolute zero temperature, the rest energy must be mainly the mass equivalent of energy for the object at rest. However, at finite temperatures for the object at rest, the rest energy must be due, at least in part, to the internal motions of the molecules, atoms, nuclei, etc. that make up the matter of the object. Keep in mind that matter is predominantly space and that the particles that make up matter undergo motions that become more violent as the temperature increases.

Note that for the condition $v \ll c$, the relativistic expression of E_{Total} in Eq. (2.37) must reduce to the classical kinetic energy, $E_K = mv^2/2$. This must be so since the classical result must always be a special case of the more general relativity result. Thus, to obtain the classical result, mass must be constant, i.e., $m_0 = m \neq m(v)$, so that Eq. (2.32) becomes

$$E_{K_0} = m_0 \int_0^v v \cdot dv = \frac{1}{2} m_0 v^2. \tag{2.38}$$

Interestingly, by introducing Eq. (2.29) into Eq. (2.37) and expanding the result in a binomial power series there results for the nonrelativistic realm of small v/c,

$$E_{Total} = mc^2 = m_0c^2(1 - v^2/c^2)^{-1/2} = m_0c^2(1 + v^2/2c^2 + \cdots) = m_0c^2 + m_0v^2/2$$

where the higher order terms become negligible. But $m_0 v^2/2$ is the nonrelativistic term for the kinetic energy E_{K_0} with the result of Eq. (2.37):

$$E_{Total} = mc^2 = m_0 c^2 + E_{K_0} = \frac{m_0 c^2}{\sqrt{1 - v^2/c^2}}. \tag{2.39}$$

This demonstrates that Eq. (2.37) applies to both the relativistic and nonrelativistic frames of reference. This is the law of *conservation of total energy* which, according to the principle of relativity, must hold in all inertial frames of reference even if the higher order terms in the expansion are no longer good approximations.

2.10 PROBLEMS

1. Two spaceships pass by each other going in the opposite direction. If they are both traveling at a velocity of $0.5c$, what is their velocity relative to each other?
   Answer: $0.80c$.

2. A river one mile across flows at an average surface velocity of 5 mi h^{-1}. How fast must a boat travel to reach the opposite shore in 6 min (see Fig. 2.1)?
   Answer: 11.2 mi h^{-1}.

3. Suppose that the earth had been subjected to an ether wind of velocity 3.0×10^6 m s^{-1} at the time Michelson–Morley performed their famous experiment. What would have been the difference between the longitudinal and transverse transit times for light in parts per million of the larger time that they would have measured?
   Answer: 50 parts per million.

4. A spaceship leaves a space station for a round trip to a distant star. An astronaut who is 35 years of age leaves his twin brother at the space station. If the spacecraft travels at an average velocity of $0.70c$ and returns to the space station after an astronaut time of 10 years, how old will be the brothers at the conclusion of the journey?
   Answers: Astronaut is 45 years of age; stay-at-home twin brother is 49 years of age.

5. A spaceship traveling in space at a velocity of $0.55c$ relative to a space station fires off a probe at a velocity of $0.60c$ relative to the spaceship. What is the velocity of the probe relative to the space station?
   Answer: $0.86c$.

6. At what velocity will a spaceship's mass be five times its rest mass?
   Answer: $0.98c$.

7. An object is known to have a density of 1000 kg m^{-3} when it is traveling in space at 500 m s^{-1}. What would be its density if its velocity were increased to $0.75c$?
   Answer: 1512 kg m^{-3}.

8. A spacecraft is moving in space at a relativistic velocity, v, as determined by the earth observers. What is the ratio of the distance observed from spacecraft-to-earth to that observed from earth-to-spacecraft if the average spacecraft velocity is $0.75c$?
   Answer: 0.66.

9. A spacecraft having a rest mass of 10^6 kg is traveling at a velocity of $0.5c$. What is its total energy in GJ (giga joules)?
   Answer: $E_{Total} = 1.04 \times 10^{14}$ GJ.

CHAPTER 3

Relativistic Rocket Mechanics

3.1 RELATIVISTIC (PROPER) MEASUREMENTS AND CALCULATIONS OF THE ASTRONAUTS

For any relativistic system, two frames of reference are required as in Fig. 3.1. For our purposes, one frame R will be called the inertial or rest frame, that of the earth. The second frame R' will be the astronaut's noninertial reference frame of the rocket that is assumed to be at constant acceleration. Postulate I of Einstein's special relativity theory given in Section 2.3 asserts that the laws of physics are valid only in an inertial reference frame, one at rest or in uniform motion. Special relativity deals with a constant acceleration of a rocket as passing continuously from one inertial frame to another. Each inertial frame occurs over a small interval of proper time during which the proper velocity undergoes a "Lorentz boost" that is always the same. In this way, special relativity theory permits velocity to increase at a constant rate with respect to the rocket's proper time.

Fig. 3.1(a) lists the quantities measured or calculated by stay-at-home scientists and engineers on Earth in rest frame R, and Fig. 3.1(b) lists the proper quantities that are measured or calculated by the astronauts in the spacecraft moving in the noninertial frame of reference, R'.

Let us first visit the astronauts in Fig. 3.1(b) assumed to be in the rocket traveling at constant acceleration in resistance-free and gravity-free space along a straight path. We assume that the astronauts can maintain a constant proper acceleration b_e by means of a calibrated accelerometer—such an accelerometer might consist of a spring onto which is attached a small mass that is caused to move along a calibrated scale as a result of the force transferred to it by the acceleration or deceleration of the rocket. We further assume that the astronauts are equipped with accurate chronometers for measurements of their proper time t_e of flight. The instantaneous proper rocket mass m_e (essentially all fuel) and the proper rate of change of fuel mass dm_e/dt_e are also assumed to be measured or calculated by the astronauts. This leaves the instantaneous proper rocket velocity v_e and the instantaneous proper distance traveled S_e to be calculated by the astronauts from measurable data. The proper propellant exhaust (efflux) velocity u_e is assumed to be known and constant, though as yet we have not considered any specific propellant for the spacecraft nor the mechanism by which it is generated.

34 RELATIVISTIC FLIGHT MECHANICS AND SPACE TRAVEL

As Observed by Earth Engineers and Scientists
(Rest Frame, R)

Earth

v = Velocity of Rocket (Measured)
b = Acceleration of Rocket (Measured)
S = Distance Traveled (Measured)
t = Time of Flight (Measured)
m = Instantaneous fuel Mass (Calculated)
dm/dt = Fuel mass change rate (Calculated)
u = Propellant exhaust velocity (Known)

(a)

As Observed by Astronauts in Rocket
(Astronaut's Non-inertial Frame, R')

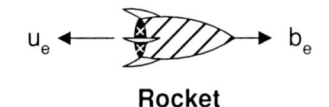

$u_e \leftarrow$ Rocket $\rightarrow b_e$

b_e = Constant proper acceleration (Measured)
t_e = Proper time of flight (Measured)
m_e = Instantaneous proper fuel mass (Measured)
dm_e/dt = Proper fuel mass change rate (Measured)
v_e = Instantaneous proper rocket velocity (Calculated)
S_e = Instantaneous proper distance traveled (Calculated)
u_e = Constant proper propellant exit velocity (Known)

(b)

FIGURE 3.1: (a) Quantities measured or calculated by earth engineers and scientists with reference to rest frame R (b) The proper quantities as measured or calculated by the astronauts in the astronaut's nonintertial frame, R', of the space vehicle

The stay-at-home engineers and scientists in Fig. 3.1(a) must rely on standard astronomical instruments and chronometers for the measurement of rocket velocity v, acceleration b, distance traveled S, and the time of flight t. The instantaneous fuel mass m and its rate of use dm/dt must be calculated by the earth observers from known facts and existing measured quantities.

From the measured proper acceleration b_e and proper time of flight t_e, the astronauts can calculate their instantaneous proper velocity v_e and instantaneous proper distance traveled S_e, both as functions of proper time. Assuming that the space flight begins with $v_e = 0$ at $t_e = 0$ relative to rest frame R, then v_e and S_e can be calculated as follows. Noting that $dv_e = b_e(t_e)dt_e$ and $dS_e = v_e(t_e)dt_e$, where b_e and v_e are functions of time t_e, there results

$$v_e = \int_0^{t_e} b_e(t_e)dt_e \tag{3.1}$$

and

$$S_e = \int_0^{t_e}\int_0^{t_e} b_e(t_e)dt_e dt_e = \int_0^{t_e} v_e(t_e)dt_e. \tag{3.2}$$

The astronauts will measure and sense the acceleration to which they are subjected as well as the passage of time indicated by their chronometers. They may check their vital signs often only to find them all to be normal. For example, their heartbeat rates will be exactly what they expected based on data taken prior to the space flight—a normal pulse rate when measured by their watches. So, according to the time dilation effect for the astronauts traveling at relativistic velocities all clocks mechanical and biological will be slowed compared to those of the stay-at-home earth observers (see Section 2.5). But the astronauts will not be aware of this fact simply by observation of their chronometers and the measurements of their heartbeat rates. Useful comparisons by direct (instantaneous) communication between the Earth observers and astronauts will not be possible.

At this point we assume that the astronauts in the spaceship of Fig. 3.1(b) are subjected to a comfortable constant proper acceleration b_e of say 10 m s^{-2} = $1g$, where g is the acceleration due to gravity. Then, applying Eqs. (3.1) and (3.2) for $b_e = const.$, the instantaneous values for proper velocity and proper distance traveled are easily determined from the classical relations:

$$v_e = b_e t_e \qquad (3.3)$$

and

$$S_e = b_e(t_e^2/2) = v_e^2/2b_e. \qquad (3.4)$$

In making the calculation for v_e by either Eq. (3.1) or (3.3), the astronauts will become aware that the proper velocity, v_e, is *not* a velocity in the Newtonian sense, and is not in violation of the *correspondence principle* (see Glossary). They discover that v_e can exceed the speed of light! To support this remarkable conclusion, an independent calculation is needed by the astronauts, one which will be considered in Section 3.2. However, before we go there, it should be pointed out that, were it possible for the astronauts to have direct instantaneous communication with the observers on earth possibly light years distant, they would learn that their measurements and calculations were in significant disagreement, a fact to be explored in Section 3.4.

3.2 INTRODUCTION TO ROCKET MECHANICS; THE ROCKET EQUATION

Consider that the rocket in Fig. 3.1(b) is traveling at relativistic speeds. To derive the *rocket equation*, we begin with Newton's generalized second law given by Eq. (2.31), and his third law: Every action is associated with an equal and opposite reaction. These laws are applied to the spaceship in Fig. 3.1(b) for the case where an instantaneous proper rocket mass m_e ejects a constant propellant mass flow rate dm_e/dt_e at a constant efflux velocity u_e (negative relative to v_e) resulting in a proper rocket acceleration dv_e/dt_e in an apposing direction to efflux ejection:

Newton's second law. Force on the rocket resulting from efflux production = $m_e(dv_e/dt_e)$.
Newton's third law. Opposing reaction in the direction of efflux ejection = $-(dm_e/dt_e)u_e$.

36 RELATIVISTIC FLIGHT MECHANICS AND SPACE TRAVEL

The rocket equation. Combining these laws yields the proper thrust, F_e, given to the rocket as

$$F_e = m_e \frac{dv_e}{dt_e} = -u_e \frac{dm_e}{dt_e}. \tag{3.5}$$

Here m_e is the instantaneous proper rocket mass (nearly all fuel mass), dm_e/dt_e is the time rate of change of m_e, u_e is the proper efflux velocity (relative to the rocket) through the plane of the nozzle, and v_e is the proper rocket velocity. The minus sign results because u_e and v_e are velocities in opposition, as indicated in Fig. 3.1(b). Equation (3.5) assumes that all (or nearly all) the rocket mass is propellant fuel mass, hence rocket borne, and that it is completely convertible to the propellant required to produce a constant proper efflux velocity, u_e, the ideal adiabatic conditions. Numerically, the *effective proper efflux velocity* is defined from Eq. (3.5) as

$$u_e = \left| \frac{F_e}{dm_e/dt_e} \right| = \frac{\text{Proper Thrust}}{\text{Proper Mass Flow Rate of Propellant}}. \tag{3.6}$$

From Eq. (3.5), the infinitesimal proper velocity change of the rocket is given by

$$dv_e = -u_e \frac{dm_e}{m_e}. \tag{3.7}$$

Setting Eq. (3.7) in integral form and assuming that the space flight begins at $v_e = 0$, there results

$$\int_0^{v_e} dv_e = -u_e \int_{m_{eo}}^{m_e} \frac{dm_e}{m_e}$$

or after integration

$$v_e = -u_e \ln \frac{m_e}{m_{eo}}. \tag{3.8}$$

Here m_{eo} and m_e are the initial and the instantaneous proper mass of the rocket, respectively. Thus, $(m_{eo} - m_e)$ is the total reaction fuel mass consumed (e.g., when m_e is the burnout mass). We now combine Eqs. (3.3) and (3.8) in the following form:

$$\frac{v_e}{u_e} = \frac{b_e t_e}{u_e} = -\ln \frac{m_e}{m_{eo}} = \ln \frac{m_{eo}}{m_e}. \tag{3.9}$$

Historically, Eq. (3.8), together with its various other forms as in Eqs. (3.9), is known as the Tsiolkovsky's rocket equation in honor of Konstantin Tsiolkovsky, the Russian physicist who derived it in the early 1900s. Today, it is referred to simply as the *rocket equation*. (See Appendix C for a detailed derivation of the ideal rocket equation as applied to any propellant.)

Except for v_e, all quantities in Eq. (3.9) are assumed to be measurable or known by the astronauts. Hence by using Eq. (3.9) and comparing results with Eqs. (3.1) and (3.2) or

Eqs. (3.3) and (3.4), the astronauts can achieve confirmation of the correctness of their measurements and calculations.

Specific impulse. Provided that a consistent set of units are used, the value for u_e is by Eq. (3.6) numerically equal to the *specific impulse* of the propellant defined as follows,

$$I_{SP} = \frac{F_e}{dm_e/dt_e} = u_e = \text{Proper Efflux Velocity}, \tag{3.10}$$

in units of force × time/mass. Again, the specific impulse given by Eq. (3.10) assumes that a consistent set of units be used so as to equate $I_{SP} = u_e$ in SI units of m s^{-1}, or in the English system, ft s^{-1}. However, in the rocket industry, it is often more convenient to use an inconsistent (mixed) set of units for specific impulse such as lbf s lbm^{-1} = s. To make a conversion from consistent to mixed units, it is customary to define specific impulse as

$$I_S = I_{SP}/g \quad \text{(in seconds, s)} \tag{3.10a}$$

where g is the acceleration due to gravity. Thus, if I_{SP} is given in SI units (m s^{-1}) then g must be $g = 9.80$ m s^{-2} and by conversion $I_S = I_{SP}/9.80$ in seconds. Similarly, if I_{SP} is given in ft s^{-1} then $g = 32.12$ ft s^{-2} and by conversion $I_S = I_{SP}/32.12$ in seconds.

The specific impulse of the efflux is often used to compare various thrusts per unit mass flow rate of propellant. For example, a common chemical propellant might have a specific impulse 4000 m s^{-1} compared to a nuclear fission (heat transfer) specific impulse of 12,000 ms^{-1}. These I_{SP} values are very small compared to the maximum obtainable $I_{SP} = c = 3 \times 10^8$ m s^{-1} for a photonic rocket to be discussed later.

Propulsive efficiency. Another quantity that proves useful for comparison purposes is the *propulsive efficiency* η_{Pe} defined as

$$\eta_{Pe} = \frac{\text{Kinetic Power of Rocket}}{\text{Kinetic Power of Rocket } + \text{ Residual Power of Efflux}} \times 100. \tag{3.11}$$

Here $F_e v_e$ is the proper kinetic power of the rocket, and by Eq. (3.10) the proper thrust is numerically $F_e = u_e(dm_e/dt_e)$. Then, assuming classical rocket mechanics $u_e \ll c$, the residual proper power of the efflux is $(dm_e/dt_e)(u_e - v_e)^2/2$ since the efflux retains a proper absolute velocity after ejection equal to $(u_e - v_e)$. Recall from Eq. (2.38) that in classical mechanics the kinetic energy is of the form $mv^2/2$. Now, introducing this information into Eq. (3.11) gives the theoretical proper propulsive efficiency ($\times 100$: in percent) as

$$\eta_P = \frac{F_e v_e}{F_e v_e + (dm_e/dt_e)(u_e - v_e)^2/2}$$
$$= \frac{u_e v_e (dm_e/dt)}{u_e v_e (dm_e/dt_e) + (dm_e/dt_e)(u_e - v_e)^2/2}$$
$$= \frac{2u_e v_e}{u_e^2 + v_e^2},$$

or finally,

$$\eta_P = \frac{2v_e/u_e}{1 + (v_e/u_e)^2} \times 100 \qquad (3.12)$$

where, as before, v_e and u_e are the proper rocket and efflux velocities, respectively. Note that the proper propulsive efficiency is valid in the range $0 \leq v_e > u_e$ and that 100% propulsive efficiency occurs under the condition $v_e/u_e = 1$, is 80% at $v_e/u_e = 2$, and drops to 10% efficiency at $v_e/u_e = 20$. Since the residual proper power of the efflux is proportional to $(u_e - v_e)^2$, $v_e > u_e \ll c$ is permitted. Remember that it is not the efflux velocity itself that propels a spacecraft but rather the reactive mechanism within the rocket that produces the efflux via Newton's third law:—Every action is associated with an equal and opposite reaction. Fig. 3.2 shows a plot of Eq. (3.12) in the range of $0 \leq v_e/u_e \leq 20$, as an example. Thus, rocket velocities in free space can exceed the efflux velocity of the propellant, but in the classical sense $u_e \ll c$.

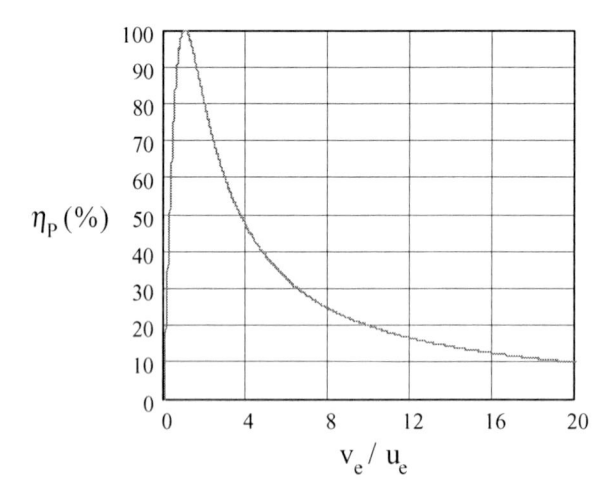

FIGURE 3.2: Plot of Eq. (3.12) giving the proper propulsive efficiency as a function of the proper rocket to efflux velocity ratio.

Not taken into account in Fig. 3.2 is the *internal efficiency* η_I of the fuel reactants. This efficiency lumps together all the efficiency factors (thrust efficiency, thermal efficiency, frozen flow efficiency, etc.) that influence the production of a directed particulate efflux in the rocket's engine or reaction chamber. The total efficiency is then given by

$$\eta_T = \eta_I \cdot \eta_P < \eta_P. \quad (3.13)$$

Considering Fig. 3.2, it follows that η_T would likely limit the rocket's velocity to $v_e < 12\, u_e$ as it relates to interstellar space travel. In conventional (chemical) rocketry it is common practice to assign 0.1% of the initial total rocket mass to payload, 11% to fuel tanks and engines, leaving 88.9% of the initial total rocket mass as fuel. Then by the rocket equation, Eq. (3.9), $v_e = u_e \ell n(100/11.1) = 2.20\, u_e$, where the burnout mass is 11.1% of the total mass. For interstellar space travel involving distances in light years, we must assume that the proportion of fuel mass relative to total rocket mass must be huge together with very large efflux velocities, u_e. As an example, suppose that the fuel mass is 99.999% of the total mass of the rocket thereby leaving 0.001% for everything else. Under this condition, the limiting rocket velocity would become $v_e = u_e \ell n(100/0.001) = 11.5\, u_e$. However, there are no known particulate propulsion systems with specific impulse magnitudes large enough to make such journeys. Such specific impulses would have to be more than 1000 times greater than current particular propellants.

Thrust to efflux-power ratio. A quantity that is useful in evaluating the relative economy of propellant materials is the theoretical proper thrust to efflux power ratio defined as

$$\frac{F_e}{P_e} = \frac{(dm_e/dt_e)u_e}{(dm_e/dt_e)u_e^2/2} = \frac{2}{u_e} = \frac{2}{I_{SP}} = \frac{\text{Thrust}}{\text{Efflux Power}} \quad (3.14)$$

where the *proper efflux power* is given by $P_e = (dm_e/dt_e)u_e^2/2$. Thus, Eq. (3.14) shows that the thrust obtainable per unit efflux power expended varies as the inverse of the specific impulse. In the SI system, the thrust to efflux power ratio is given in s m^{-1} units. As a typical example, the peak specific impulse I_{SP} of a common fuel-oxidant propellant (liquid H$_2$–liquid oxygen) is about 4000 m s^{-1}. For this propellant, the thrust to efflux power ratio becomes of the order of 5×10^{-4} s m^{-1} equal to 8.4×10^{-2} lbf hp^{-1}. The latter quantity indicates that this particular propellant can develop about 8.4 pounds of thrust for each 100 hp of fuel power, a relatively high value for chemical propellants. The ratio of the theoretical thrust to efflux power ratio given by Eq. (3.14) to the actual thrust to efflux power ratio $\times 100$ is called the *thrust efficiency*, and is a useful measure of the economy of a propellant.

3.3 THE PHOTON ROCKET

Now, to greatly simplify the relativistic flight mechanics, we assume that the rocket in Fig. 3.1 is a *photon rocket* whose propellant velocity is that of the speed of light c. Appendix C provides a detailed derivation of the *photon rocket equation*. The results are shown in the forms

$$\frac{v_e}{c} = -\ell n \frac{m_e}{m_{eo}} = \ell n \frac{m_{eo}}{m_e} \qquad (3.15)$$

or

$$\frac{m_{eo}}{m_e} = e^{v_e/c}. \qquad (3.15a)$$

Here v_e/c is called the *proper Einstein number* equal to the ratio of proper flight velocity to the exhaust velocity, c. Hence, the proper Einstein number is a normalized proper velocity. Just as the *mach number* is used to express supersonic flight velocities (e.g., mach 1, mach 2, etc.), the proper Einstein number has a similar meaning with regard to the velocity of light. For example a "trekkie" might refer to proper Einstein numbers equal to or in excess of the speed of light in terms of warp number (warp 1, warp 2, etc.). Thus, knowing the initial proper mass m_{eo} and monitoring the instantaneous proper mass, the astronauts can plot their progress in terms of their proper Einstein number according to Eq. (3.15). Conversely, the astronauts can plan their space trip in terms of the proper Einstein number they wish to achieve and adjust the initial to instantaneous proper mass ratio accordingly.

Shown in Fig. 3.3 is a semilog plot of Eq. (3.15a) where the ratio of the initial proper mass to the instantaneous proper mass is given as a function of proper Einstein number, v_e/c. Fig. 3.3 demonstrates the very large initial proper mass compared to the instantaneous proper mass required to maintain the proper Einstein number indicated. For example, to achieve a proper Einstein number of 7 would require that $m_{eo} \cong 10^3 \, m_e$, meaning that the initial proper mass must be about 1000 times that of the instantaneous proper mass. Or at a proper Einstein number of 14 the instantaneous proper mass would have to be $m_{eo} \cong 10^6 \, m_e$. This matter will be revisited later in this chapter and in Chapter 4.

In applying Eq. (3.15) or Fig. 3.3 to their progress in space flight, the astronauts will be made aware that their perceived speed v_e can exceed the speed of light by more than an order of magnitude. Of course, v_e is not a velocity in the classical sense but is a consequence of the photonic propulsion system and the assumptions made with regard to rocket mass and fuel mass given earlier. As we shall see in later sections, the proper Einstein number will be highly useful in dealing with a variety of relativistic parameters.

During the relativistic flight process, it is assumed that all or nearly all the rocket mass is convertible to photon energy by means of Einstein's energy/mass relation (2.37) expressed as

$$E_e = m_e c^2 = \hbar f = \hbar c/\lambda \qquad (3.16)$$

where $\hbar = 6.625 \times 10^{-34}$ J s is Plank's constant, f is the average frequency of photonic radiation, and λ is the average wavelength of the photon radiation. Thus, we assume that the instantaneous proper mass, as observed by the astronauts, is completely (or nearly completely)

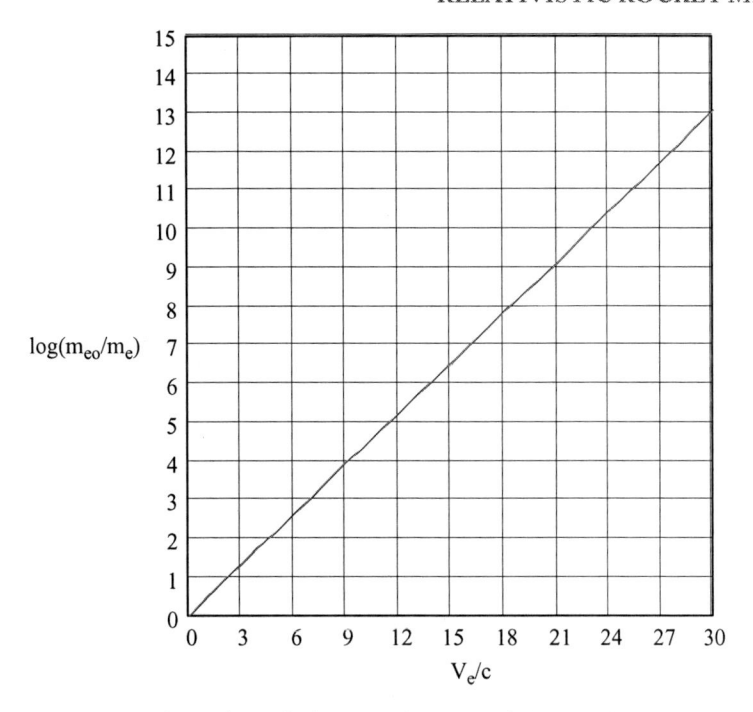

FIGURE 3.3: Semilog plot of Eq. (3.15a) showing the ratio of the initial proper mass to the instantaneous proper mass as a function of the proper Einstein number

transformed into photon energy according to Eq. (3.16) where the mass equivalent of energy is expressed as

$$m_e = \hbar/\lambda c. \quad (3.17)$$

Thrust and specific impulse of a photon rocket. At first thought, it seems unnatural to expect that photonic radiation is capable of producing a thrust at all via Newton's third law let alone one sufficient to propel a spacecraft to relativistic velocities. Remember that an ejected photon from the "nozzle" of a spacecraft has no rest mass in the sense that a particulate efflux particle has rest mass. Of course, most of us have seen a comet's tail always oriented in opposition to the sun's direction regardless of the comet's position relative to the sun. This is due to the so-called "radiation pressure" of the photons that interact with the particles on the comet. This radiation pressure results from the dual wave/particle nature of radiation quanta. The photon theory, which treats quanta as having the nature of particles, is used to explain the photoelectric effect where electrons are ejected from a metal surface under high energy photon bombardment.

The concept of a radiation pressure (thrust) can best be understood by combining Eqs. (2.31), (3.10), and (3.16) as applied to photons to give

$$F_e = \frac{dp_e}{dt_e} = c\frac{dm_e}{dt_e} = \left(\frac{dE_e/dt_e}{c^2}\right)c = \frac{dE_e/dt_e}{c} = \frac{P_e}{c} \quad (3.18)$$

where F_e is the proper thrust produced by the internal reaction resulting in the photon propellant, p_e is the proper photon momentum, and $P_e = dE_e/dt_e$ is the proper efflux power of the photon propellant. From Eqs. (3.10) and (3.18), the photon *specific impulse* I_{SP} is

$$I_{SP} = F_e/(dm_e/dt_e) = u_e = c = 3.00 \times 10^8 \text{ m s}^{-1}. \quad (3.19)$$

in consistent units. In mixed SI units, the specific impulse of a photonic propellant is $I_S = I_{SP}/g$ or $I_S = 3.00 \times 10^8/9.80 = 3.06 \times 10^7$ s, where $g = 9.80$ m s^{-1}. Since a photon efflux has no rest mass, g is regarded as simply a numerical conversion factor.

Propulsive efficiency of a photon rocket. Acknowledging that now $u_e = c$ for the case of a photon rocket, it is necessary to recalculate the propulsive efficiency. To do this, it must be recalled that the relativistic kinetic energy is of the form mc^2, not $mv^2/2$ as in the classical case. Also, a photon does not have rest mass, but has what is improperly termed "relativistic mass" which must be expressed in terms of Einstein's mass–energy transfer expression, $E = mc^2$. Then, since by Einstein's Postulate II, the velocity of light must be constant independent of the velocity of the source of the photons, we must conclude that after ejection from the rocket's nozzle plane a photon retains its absolute velocity, c, relative to the rocket. Thus, from Eq. (3.18), the residual power lost by the photon efflux becomes $(dm_e/dt_e)c^2$, and the proper thrust F_e produced by the photon propellant is $(dm_e/dt_e)c$. Introducing this information into Eq. (3.12) and taking $F_e v_e = cv_e(dm_e/dt_e)$ as the proper kinetic power of the photon rocket, there results

$$\eta_{P,\ Photon} = \frac{F_e v_e}{F_e v_e + (dm_e/dt_e)c^2} \times 100$$

$$= \frac{cv_e(dm_e/dt_e)}{cv_e(dm_e/dt_e) + c^2(dm_e/dt_e)} \times 100$$

$$= \frac{cv_e}{cv_e + c^2} \times 100$$

or simply

$$\eta_{P,\ Photon} = \frac{v_e/c}{1 + v_e/c} \times 100. \quad (3.20)$$

Fig. 3.4 shows a plot of the propulsion efficiency for a photon rocket expressed by Eq. (3.20). Note the contrast between the propulsion efficiency of a nonrelativistic rocket in Fig. 3.2 and that of a relativistic photon rocket of Fig. 3.4. This is due to the difference in the proper

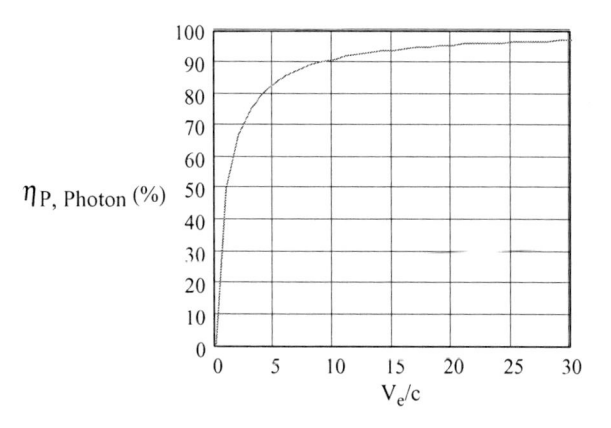

FIGURE 3.4: Plot of Eq. (3.20) giving the proper propulsive efficiency for a photon rocket as a function of proper Einstein number

absolute velocity of the efflux (relative to the rocket) between the two cases, $(u_e - v_e)$ for the nonrelativistic case, and c for the photon rocket, which remains constant independent of the rocket's velocity. Whereas the propulsive efficiency for a nonrelativistic rocket reaches 100% when $v_e/u_e = 1$ and drops off thereafter with increasing v_e/u_e, the case of the relativistic photon rocket is quite different. Fig. 3.4 shows that at $v_e/c = 1$ the efficiency is 50% and that 100% efficiency is approached only as v_e/c becomes very large. As we move through the remaining properties of the photon rocket, we will have cause to refer to the propulsive efficiency of the photon rocket.

Following Eq. (3.13), the overall efficiency must take into account the internal efficiency, η_I, which lumps together all efficiency factors including thrust efficiency, thermal efficiency, frozen flow efficiency, etc., that affect the directed efflux from a photon rocket. Thus, the overall efficiency for a photon rocket could be given by

$$\eta_T = \eta_I \cdot \eta_{P,\,Photon} < \eta_{P,\,Photon}. \tag{3.21}$$

It is likely that η_I may turn out to be the dominate efficiency factor in Eq. (3.21).

Photonic thrust to efflux power ratio. The thrust to efflux power ratio for particulate propellants was given by Eq. (3.14). Now for comparison purposes, the photonic thrust to efflux power ratio for a photon rocket as obtained from Eq. (3.18) is given by

$$\begin{aligned}\frac{F_e}{P_e} &= \frac{1}{c} = 3.33 \times 10^{-9} \text{ s m}^{-1} \\ &= 3.33 \times 10^{-6} \text{ N kW}^{-1} \\ &= 0.560 \times 10^{-6} \text{ lbf hp}^{-1}.\end{aligned} \tag{3.22}$$

Since velocity c is the highest efflux velocity possible, it follows that the thrust–efflux power ratio must be the lowest possible. The last value for $F_e/P_e = 0.560 \times 10^{-6}$ lbf hp^{-1} in Eq. (3.22) indicates that 0.56 pounds of thrust results for each one million horse power generated by the photonic efflux, or alternatively, a force of 3.33 N results from each GW of photonic power generated. Thus, a space vehicle of massive proportions is required to propel a spaceship to relativistic velocities over any reasonable time frame. This is quite consistent with Fig. 3.3. The propellant reactants, as yet unspecified, must occupy nearly all of the rocket's mass and be completely converted to usable photonic energy according to Eq. (2.37). This was the assumption made early on in the developments required of any rocket that is designed to travel at relativistic velocities. However, this requirement may need to be modified as we explore various other possibilities in Chapter 6.

3.4 RELATIVITY OF VELOCITY, TIME, ACCELERATION, AND DISTANCE

In this section, we will explore the theoretical relationships between measurements and calculations made by the astronauts and those made by the earth observers. Were it possible for the astronauts to communicate the status of their fuel consumption m_e/m_{eo} to the earth observers, it would become apparent that considerable but necessary discrepancies exist between astronaut and earth observations and calculations.

We begin by establishing a new set of transformation equations between the rest frame R (earth) and an inertial frame R' for a rocket moving at uniform relativistic velocity, v. Referring to Fig. 3.1, the transformation equations for m, dm, and u'_x are given as follows,

$$
\begin{aligned}
m &= \frac{m_e}{\sqrt{1 - v^2/c^2}} \\
dm &= d\left(\frac{m_e}{\sqrt{1 - v^2/c^2}}\right) \\
u'_x &= \frac{v - u_e}{1 - u_e v/c^2}
\end{aligned}
\qquad (3.23)
$$

where the transformation for u'_x expresses the summation of velocities as given in Eq. (2.25). The term $(v - u_e)$ is due to velocity vectors in opposition, as shown in Fig. 3.1. The normalized velocity, v/c, is called the *Einstein number* equal to the ratio of the rocket velocity as observed from the earth to the velocity of light, c. As always, subscript e denotes the proper quantities as observed by the astronauts.

The conservation of momentum is now recast into differential form as follows:

$$d(mv) = m\,dv + v\,dm = u'_x\,dm.$$

RELATIVISTIC ROCKET MECHANICS

Then solving for mdv gives

$$mdv = (u'_x - v)dm$$

$$= \left(\frac{v - u_e}{1 - u_e v/c^2} - v\right) dm \qquad (3.24)$$

$$= -u_e \left(\frac{1 - v^2/c^2}{1 - u_e v/c^2}\right) dm$$

after clearing fractions and factoring. We now solve for dm/m with the results

$$\frac{dm}{m} = -\left(\frac{1 - u_e v/c^2}{u_e(1 - v^2/c^2)}\right) dv. \qquad (3.25)$$

But introducing the transformations for dm and m in Eqs. (3.23) gives for dm/m the following:

$$\frac{dm}{m} = \frac{\sqrt{1 - v^2/c^2}}{m_e} \cdot d\left(\frac{m_e}{\sqrt{1 - v^2/c^2}}\right). \qquad (3.26)$$

To proceed, it is necessary to expand the expression for $d(m_e/\sqrt{1 - v^2/c^2})$ as follows,

$$d\left(\frac{m_e}{\sqrt{1 - v^2/c^2}}\right) = \frac{dm_e}{\sqrt{1 - v^2/c^2}} - m_e d(1 - v^2/c^2)^{1/2}$$

$$= \frac{dm_e}{\sqrt{1 - v^2/c^2}} - m_e \frac{(1 - v^2/c^2)^{-3/2} \cdot (-2v/c^2)}{2} dv$$

or

$$d\left(\frac{m_e}{\sqrt{1 - v^2/c^2}}\right) = \frac{dm_e}{\sqrt{1 - v^2/c^2}} + \frac{m_e(v/c^2)}{(1 - v^2/c^2)^{3/2}} dv. \qquad (3.27)$$

Introduction of the resultant expression in Eqs. (3.27) into Eq. (3.26) leads to the result

$$\frac{dm}{m} = \frac{\sqrt{1 - v^2/c^2}}{m_e} \cdot \left[\frac{dm_e}{\sqrt{1 - v^2/c^2}} + \frac{m_e(v/c^2)}{(1 - v^2/c^2)^{3/2}} dv \cdot \left(\frac{u_e}{u_e}\right)\right]$$

$$= \frac{dm_e}{m_e} + \frac{u_e v/c^2}{u_e(1 - v^2/c^2)} dv. \qquad (3.28)$$

Equating (3.28) with (3.25) yields

$$\frac{dm}{m} = \frac{dm_e}{m_e} + \frac{u_e v/c^2}{u_e(1 - v^2/c^2)} dv = -\left(\frac{1 - u_e v/c^2}{u_e(1 - v^2/c^2)}\right) dv.$$

Then, solving for dm_e/m_e and simplifying gives the result

$$\frac{dm_e}{m_e} = -\frac{dv}{u_e(1-v^2/c^2)}. \quad (3.29)$$

The final series of steps requires that Eq. (3.29) be integrated under two assumptions: (1) the space flight of the rocket begins at rest with respect to the rest frame R, ($v = 0$); (2) the initial rocket mass is the proper mass m_{eo}, and a final proper mass is its instantaneous value m_e. Carrying out this integration assuming that u_e is constant and adding a little algebra yields the following important result,

$$\int_{m_{eo}}^{m_e} \frac{dm_e}{m_e} = -\frac{1}{u_e} \int_0^v \frac{dv}{(1-v^2/c^2)}$$

or

$$\ell n \left(\frac{m_e}{m_{eo}}\right) = -\frac{c}{2u_e} \cdot \ell n \left(\frac{1+v/c}{1-v/c}\right) = \ell n \left(\frac{1-v/c}{1+v/c}\right)^{c/2u_e}. \quad (3.30)$$

Then taking the antilog of both sides gives the *relativistic rocket equation* relating the mass ratio to the Einstein number—all in agreement with the correspondence principle (see Glossary),

$$\frac{m_e}{m_{eo}} = \left[\frac{1-v/c}{1+v/c}\right]^{c/2u_e}. \quad (3.30a)$$

Thus, Eq. (3.30a) yields the results for the *classical rocket equation*, Eq. (C.8) in Appendix C, when $v < 0.25c$ for which $v_e = v$ and under the condition $I_{SP} = c$.

Now, solving for v/c there results another form of the *ideal relativistic rocket equation* given by

$$\frac{v}{c} = \frac{1 - (m_e/m_{eo})^{2u_e/c}}{1 + (m_e/m_{eo})^{2u_e/c}}. \quad (3.31)$$

We can now express the relationship between Einstein number (v/c) measured by the earth observers in terms of the proper Einstein number (v_e/c) which is calculated by the astronauts. This can be done in either of two ways. We can introduce the antilog of Eq. (3.9) into Eq. (3.31) or, introduce the antilog of Eq. (3.15) into Eq. (3.31) with c replacing u_e. Thus, the two options are given as option (A) or option (B), respectively, as follows:

Option (A) $\left(\frac{m_e}{m_{eo}}\right) = e^{-v_e/u_e}$ or Option (B) $\left(\frac{m_e}{m_{eo}}\right) = e^{-v_e/c}$ (c replaces u_e).

When either option (A) or option (B) is introduced into Eq. (3.31) the results are the same

$$\frac{v}{c} = \left[\frac{1-e^{-2v_e/c}}{1+e^{-2v_e/c}}\right] = \tanh\left(\frac{v_e}{c}\right). \quad (3.32)$$

This shows that Eq. (3.32) applies to either a particulate or photon propellant equally well.

RELATIVISTIC ROCKET MECHANICS   47

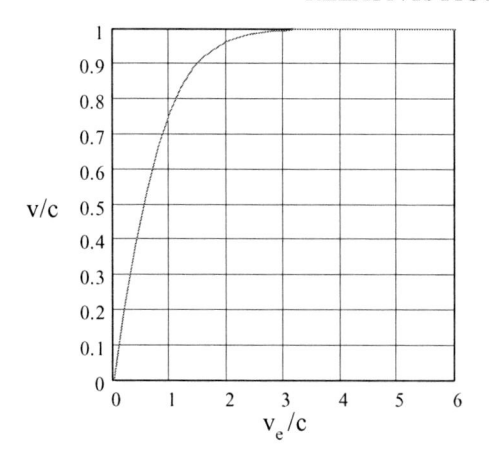

FIGURE 3.5: Einstein number v/c as a function of the proper Einstein number v_e/c expressed in Eq. (3.32)

Shown in Fig. 3.5 is a plot of Eq. (3.32) showing the earth-measured v/c as a function of v_e/c calculated by the astronauts by using Eq. (3.3) or Eq. (3.9) with $u_e = c$. As examples, the details show that $v/c = 0.762$ when $v_e/c = 1$, $v/c = 0.995$ when $v_e/c = 3.0$, and $v/c = 0.999$ at $v_e/c = 4.0$. Of course, v/c never equals unity but only approaches it asymptotically. Thus, as the proper Einstein number $v_e/c \to \infty$ the Einstein number $v/c \to 1$. Shown in Table 3.1 are representative values of v/c as a function of v_e/c taken from an expansion of Fig. 3.5.

By squaring Eq. (3.32) there results

$$(v/c)^2 = \tanh^2(v_e/c)$$

such that $1 - (v^2/c^2) = 1 - \tanh^2(v_e/c) = \text{sech}^2(v_e/c)$. Then, from Eq. (2.11),

$$\frac{1}{\sqrt{1 - v^2/c^2}} = \frac{1}{\sqrt{\text{sech}^2(v_e/c)}} = \cosh(v_e/c),$$

TABLE 3.1: Representative Values of the Einstein Number in Eq. (3.32) Observed from Earth as a Function of the Proper Einstein Number Calculated by the Astronauts

v_e/c	0.1	0.2	0.4	0.6	1.0	2.0	4.0	6.0
v/c	0.1	0.197	0.380	0.537	0.762	0.964	0.999	0.99999

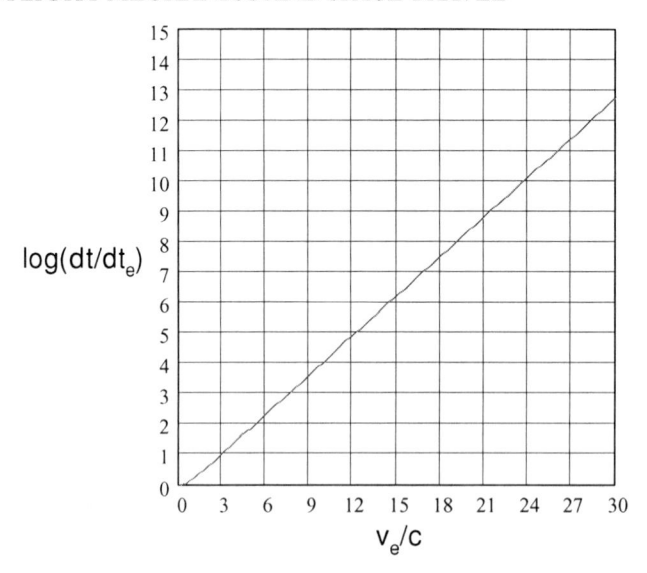

48 RELATIVISTIC FLIGHT MECHANICS AND SPACE TRAVEL

FIGURE 3.6: Semilog of time of flight ratio as a function of the proper Einstein number according Eq. (3.33)

there results

$$\frac{dt}{dt_e} = \cosh\left(\frac{v_e}{c}\right). \qquad (3.33)$$

Equation (3.33) gives the elemental time of flight ratio of the earth observer's measurement to that of the astronaut's measurement as a function of the proper Einstein number, v_e/c, or $v_e(t_e)/c$ if v_e is a function of t_e. Shown in Fig. 3.6 is a semilog plot of the time of flight ratio as a function of v_e/c according to Eq. (3.33). This plot provides the flight time (e.g., to a star) as perceived by the earth observers versus that measured by the astronauts. As an example, at a proper Einstein number of $v_e/c = 18$, $dt/dt_e = 3.28 \times 10^7$. This means that at a proper rocket velocity of $v_e = 18c$, one second of astronaut flight time is registered as one year earth time. Or at a proper rocket velocity of $v_e = 12c$, one second of astronaut time would amount to about one day of earth time. Then if $v_e = 3c$, one year of astronaut time would be about 10 years earth time, etc.

The ratio of earth-observed rocket acceleration to that measured by the astronauts can be calculated as a function of the proper Einstein number as follows:

$$\frac{b}{b_e} = \frac{dv/dt}{dv_e/dt_e} = \frac{dv}{dv_e}\left(\frac{dt}{dt_e}\right)^{-1}. \qquad (3.34)$$

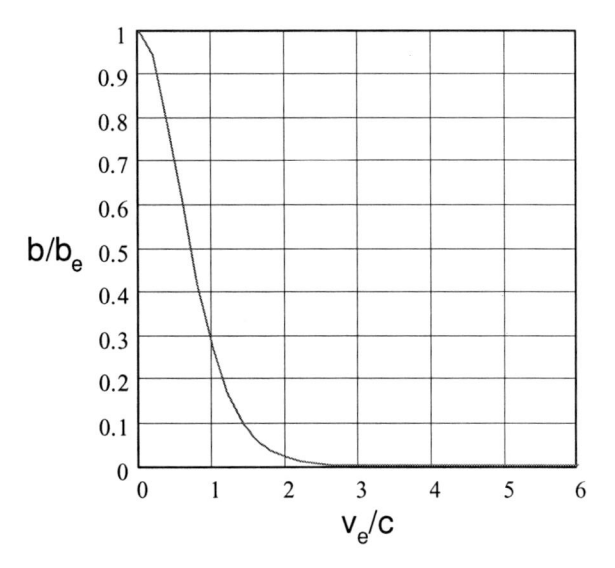

FIGURE 3.7: Acceleration ratio for earth observer versus astronauts as a function of proper Einstein number

Differentiating Eq. (3.32) gives

$$dv/dv_e = \text{sec}\,h^2(v_e/c) = \cosh^{-2}(v_e/c).$$

Then, by introducing this expression into Eq. (3.34) along with Eq. (3.33) we have the result

$$\frac{b}{b_e} = \cosh^{-3}\left(\frac{v_e}{c}\right). \tag{3.35}$$

Fig. 3.7 gives a plot of the acceleration ratio as a function of proper Einstein number, in agreement with Eq. (3.35). Here, it is clear that the acceleration b, as observed from the earth, falls off rapidly below the proper value b_e measured by the astronauts in the spacecraft. For example, at $v_e/c = 3$ the acceleration ratio is about $b/b_e = 1 \times 10^{-3}$ and at $v_e/c = 6$, $b/b_e = 1.2 \times 10^{-7}$. All of this is expected and consistent with Fig. 3.5. As v/c approaches unity (but never reaches it), b/b_e approaches zero (but never reaches it). Remember that as the velocity observed from the earth $v \to c$, the mass of the spacecraft $m \to \infty$ as does the total energy $E_{Total} \to \infty$, according to Eq. (2.39).

Still remaining is the ratio of the elemental distance (e.g., to a star) measured from the earth to the elemental proper value calculated by the astronauts. This is easily determined by

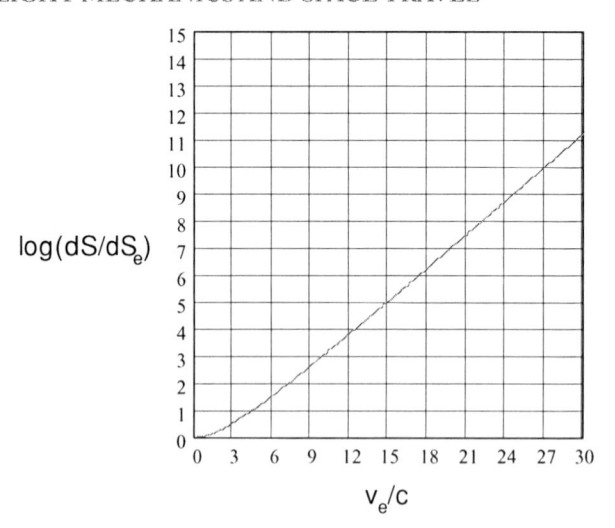

50 RELATIVISTIC FLIGHT MECHANICS AND SPACE TRAVEL

FIGURE 3.8: Distance ratio as a function of proper Einsetin number as required by Eq. (3.36)

using Eqs. (2.12), (3.32), and (3.33) as follows,

$$\frac{dS}{dS_e} = \frac{v/c}{v_e/c} \cdot \frac{dt}{dt_e} = \frac{\tanh(v_e/c)}{v_e/c} \cdot \cosh(v_e/c)$$

or

$$\frac{dS}{dS_e} = \frac{\sinh(v_e/c)}{v_e/c}. \qquad (3.36)$$

Shown in Fig. 3.8 is a semilog plot of the elemental distance ratio dS/dS_e as a function of v_e/c. The meaning of Eq. (3.36) as represented in this plot is as follows. At a proper Einstein number of $v_e/c = 18$, $dS/dS_e \cong 2 \times 10^6$. Thus, at an instantaneous proper velocity $v_e = 18c$, any distance traveled to the star viewed by the observer on earth would be about two million times that calculated by the astronauts by using Eq. (3.4). Or at an instantaneous proper velocity of $v_e \cong 10c$, the distance to the star measured by the earth observer would be about 1100 times that calculated by the astronauts. However, at a proper velocity of $v_e = 27c$, each element viewed by the earth observer would be larger by a factor of 10^{10} than that calculated by the astronauts, i.e., $dS = 10^{10} dS_e$, an astronomical difference. A few representative values of the elemental distance ratio as a function of proper Einstein number v_e/c are provide in Table 3.2. Here, the exponential increase of the distance ratio as a function of v_e/c is also dramatically displayed but now by values from an expansion of Fig. 3.8.

Equation (3.36) and Fig. 3.8 do not, however, present the complete story of distance measured by the earth observer versus that determined by the astronauts. Let us engage in another thought experiment to explore what the astronauts would be able to determine from

TABLE 3.2: Representative Values of the Elemental Distance Ratio as a Function of Proper Einstein Number, v_e/c

v_e/c	1	2	3	5	7	10	15	20	25
$(dS)/(dS_e)$	1.175	1.81	3.34	14.84	78.3	1.1×10^3	1.1×10^5	1.2×10^7	1.4×10^9

their rear window's view of the earth as in Fig. 2.6(b). We assume that the astronauts are in a spacecraft accelerating at relativistic velocities and are equipped with the same conventional astronomical instruments as those of the observers on earth. Remember that direct instantaneous communication between the spacecraft and earth is not possible. But if it were possible, let the quantities to be evaluated from the spacecraft's window be designated by a superscript *. Thus, the quantities to be evaluated are t_e^*, b_e^*, v_e^*, and S_e^*. Because of the issue of synchronicity, the time measurements t_e^* must be the proper time t_e, hence $t_e^* = t_e$. Then, since it makes no difference whether the earth at rest views the spacecraft in motion or vice versa, we conclude that $b_e^* = b \neq b_e$ and $v_e^* = v \neq v_e$. Thus, the meaning of the above is that the spacecraft-to-earth measured quantities, b_e^* and v_e^*, are the same as those measured from the earth to the spacecraft, b and v, but $t_e^* = t_e$.

Now our attention must center on the distance, S_e^*, from the spacecraft to earth determined by the astronauts as in Fig. 2.6(b). From Eq. (2.12) we can write $dS_e^* = v_e^* dt_e^* = v dt_e$ and $dS = v dt$ for the elemental distances measured from the spacecraft window to earth and that from the earth to spacecraft, respectively (see Fig. 2.6). Taking the ratio of these distances gives

$$\frac{dS}{dS_e^*} = \frac{v dt}{v dt_e} = \frac{dt}{dt_e} = \frac{1}{\sqrt{1 - v^2/c^2}} = \cosh\left(\frac{v_e}{c}\right) \quad (3.37)$$

which is the same as Eq. (3.33) and is given by the semilog plot in Fig. 3.6. Thus, the distance to earth as viewed from the spacecraft moving at relativistic velocity, v, is $dS_e^* = dS\sqrt{1 - v^2/c^2}$ and is contracted by factor $\sqrt{1 - v^2/c^2}$ relative to the distance to the spacecraft as viewed from the earth at rest. Note that this is the same as that given by Eq. (2.14). Equation (3.36), on the other hand, is quite different since the distance ratio is relative to a distant star, not between the earth and spacecraft (again, refer to Fig. 2.6). Remember that dS_e must be calculated by the astronauts from the measured proper time of flight t_e and the constant proper acceleration, b_e.

An interesting and useful development arises from Eqs. (3.36) and (3.37). Taking the ratio of these two equations leads to the following result:

$$\frac{dS/dS_e}{dS/dS_e^*} = \frac{dS_e^*}{dS_e} = \frac{\sinh(v_e/c)/(v_e/c)}{\cosh(v_e/c)} = \frac{\tanh(v_e/c)}{v_e/c} = \frac{v/c}{v_e/c} = \frac{v}{v_e}$$

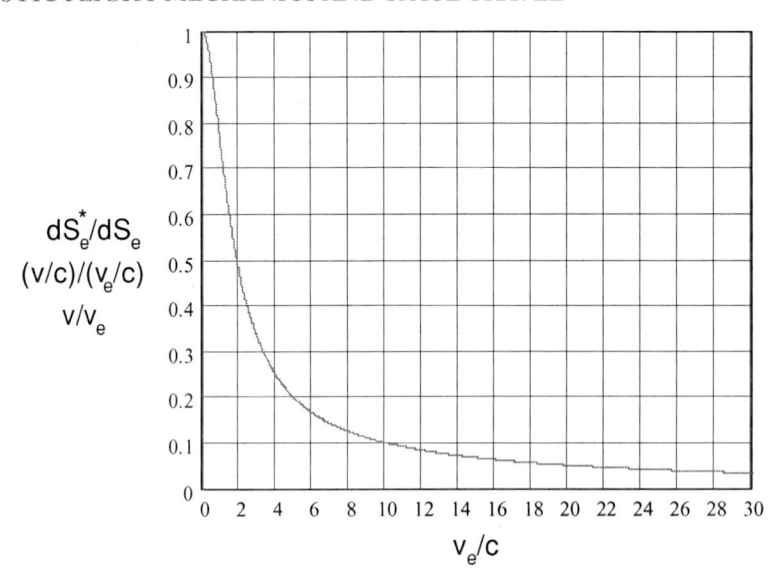

52 RELATIVISTIC FLIGHT MECHANICS AND SPACE TRAVEL

FIGURE 3.9: Plot of Eq. (3.38) showing distance, Einstein number and velocity ratios as a function of proper Einstein number

or simply

$$\frac{dS_e^*}{dS_e} = \frac{v/c}{v_e/c} = \frac{v}{v_e} = \frac{\tanh(v_e/c)}{v_e/c}. \qquad (3.38)$$

Shown in Fig. 3.9 is a graph of Eq. (3.38) where the quantities dS_e^*/dS_e, $(v/c)/(v_e/c)$ and v/v_e are plotted as a function of proper Einstein number in the range $0 \leq (v_e/c) \leq 30$. A cursory check of the Einstein number and proper Einstein number values in Fig. 3.9 reveals that they are in agreement with those obtained from Fig. 3.5 and Table 3.1.

The significance of Eq. (3.38) and Fig. 3.9 is that they show the ratio of distances measured from the window of the spacecraft to those calculated from Eq. (3.2) or (3.4) by the astronauts as a function of proper Einstein number, v_e/c. Since $v_e^* = v$, Fig. 3.9 also gives the ratio of velocities measured from the spacecraft window to those calculated from Eq. (3.1) or (3.3) by the astronauts all as a function of proper Einstein number. Recall that the calculated distances and velocities are obtained from either Eq. (3.2) or (3.4) for distances and from Eq. (3.1) or (3.3) for velocity depending on whether or not b_e is a function of time. The Einstein number ratios follow from their respective velocities. Table 3.3 shows a few sample readings from an expansion of Fig. 3.9.

From Table 3.3 it is clear that at very low proper velocities, $v_e/c < 0.2$, the earth readings will agree with those calculated by the astronauts, thus $v = v_e$ indicating a nonrelativistic range. Also at lower proper velocities, the distances measured and calculated within the spacecraft

RELATIVISTIC ROCKET MECHANICS

TABLE 3.3: Astronaut-Measured to Astronaut-Calculated Distance Ratios dS_e^*/dS_e, and Earth-Measured to Astronaut-Calculated Proper Velocity Ratios v/v_e are all Given as a Function of a Few Selected Proper Einstein Numbers, v_e/c

v_e/c	< 0.02	0.50	1.00	2.00	4.00	5.00	10	20	25
dS_e^*/dS_e, v/v_e	1.00	0.924	0.762	0.482	0.25	0.20	0.10	0.05	0.04

will agree, hence $dS_e^* = dS_e$. However, at higher proper velocities these ratios begin to change. For example, at $v_e/c = 1$, $v = 0.762v_e = 0.762c$, and at $v_e/c = 2$, $v = 0.482v_e = 0.964c$, etc. Then for even higher proper velocities, the earth-measured velocity of the spacecraft approaches c asymptotically but never reaches it exactly. Also, at higher proper velocities the measured distances dS_e^* become an increasingly smaller fraction of the calculated values, dS_e as seen from Table 3.3.

For comparison purposes, Fig. 3.10 shows the instantaneous slope of Fig. 3.5 as a function of the Proper Einstein number. Thus, the rate of change of Einstein number with the proper Einstein number is plotted as a function of v_e/c. Clearly, this rate of change of Einstein number approaches zero asymptotically with v_e/c but is essentially

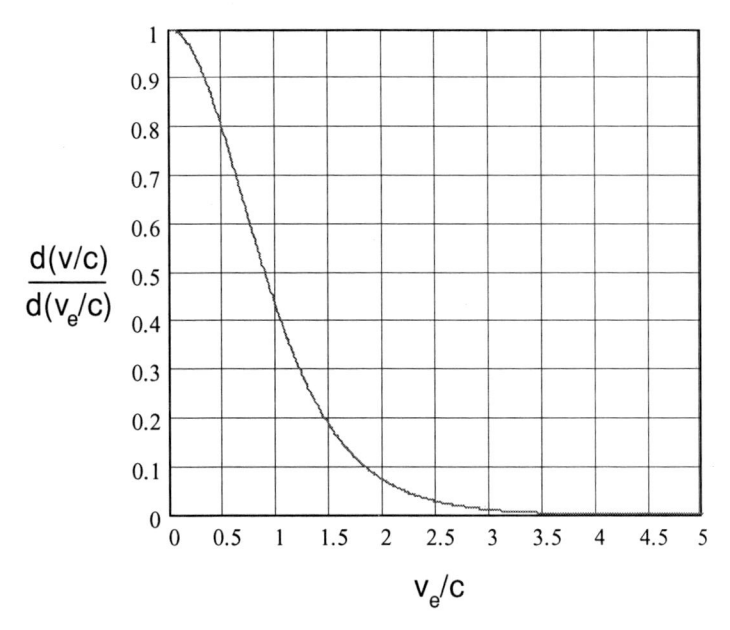

FIGURE 3.10: Plot of the slope of the curve in Fig. 3.5 as a function of Proper Einstein number, hence the rate of change of Einstein number with proper Einstein number as a function of the latter

zero for $v_e/c > 4$. This is expected from Fig. 3.5 since v approaches c asymptotically for $v_e/c > 4$.

The curve in Fig. 3.10 passes through an inflection point at $v_e/c = 0.6585$. This is easily determined from the third derivative of Eq. (3.32) set to zero (i.e., from the second derivative of $[d(v/c)/d(v_e/c)]$ set to zero). Here, $d(v/c)/d(v_e/c) = 1 - \tanh^2(v_e/c) = \sec h^2(v_e/c)$ is the first derivative of (3.32) with respect to (v_e/c):

$$\frac{d^3}{d(v_e/c)^3} \tanh(v_e/c) = 2 \cdot \frac{2\cosh^2(v_e/c) - 3}{\cosh^4(v_e/c)} = 0$$

or

$$v_e/c = \cosh^{-1}\left(\sqrt{3/2}\right) = 0.6585$$

In comparison, an inflection point occurs in Fig. 3.9 at $v_e/c = 1.60$. This added information regarding inflection points is of little interest except to show where the maximum rate of change (maximum slope) occurs for a given ratio.

3.5 ENERGY REQUIREMENTS FOR RELATIVISTIC FLIGHT

In Section 3.2, we defined m_{eo} and m_e as the initial and instantaneous mass of the rocket, respectively, such that their difference $(m_{eo} - m_e)$ is the fuel mass consumed in propelling the spacecraft; a fuel mass that is essentially the same as the instantaneous mass of the rocket. From Eq. (2.39), the average energy expended by the consumption of rocket fuel is subject to relativistic mass dilation given by

$$E_{Total} = mc^2 = \frac{(m_{eo} - m_e)c^2}{\sqrt{1 - (v^2/c^2)_{Avg}}}. \qquad (3.39)$$

But from Eq. (3.15a)

$$m_{eo} = m_e e^{(v_e/c)_{Avg}}$$

which, when introduced into Eq. (3.39) and applying Eq. (3.37), gives the average proper energy

$$E_e = \frac{m_e c^2 (e^{(v_e/c)_{Avg}} - 1)}{\sqrt{1 - (v^2/c^2)_{Avg}}} = m_e c^2 (e^{(v_e/c)_{Avg}} - 1) \cdot \cosh(v_e/c)_{Avg}. \qquad (3.40)$$

By dividing E_e by m_e there results the expression for the average proper *specific energy* given by

$$\varepsilon_e = \frac{E_e}{m_e} = c^2 (e^{(v_e/c)_{Avg}} - 1) \cdot \cosh(v_e/c)_{Avg} \qquad (3.41)$$

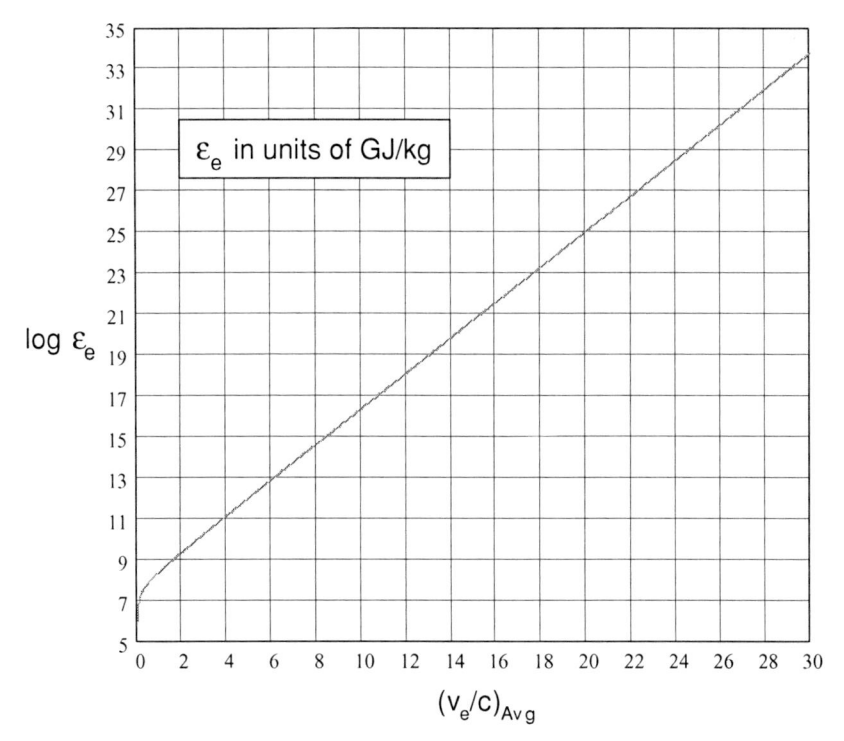

FIGURE 3.11: Semilog plot of average proper specific energy as a function of average proper Einstein number required to propell a rocket to relativistic velocities

measured in SI units of J kg^{-1} = (m^2 s^{-2}) = N m kg^{-1}. Thus, proper specific energy is the instantaneous proper energy of the rocket per unit proper mass of the rocket which is mostly fuel mass. Shown in Fig. 3.11 is a semilog plot of the average proper specific energy of Eq. (3.41) as a function of $(v_e/c)_{Avg}$ now expressed as

$$\varepsilon_e = 9.0 \times 10^7 (e^{(v_e/c)_{Avg}} - 1) \cdot \cosh(v_e/c)_{Avg} \qquad (3.42)$$

in units of GJ kg^{-1} (giga joules per kilogram of fuel).

For the case of $v_e \ll c$, $(e^{(v_e/c)_{Avg}} - 1) \cdot \cosh(v_e/c)_{Avg} \to (v_e/c)_{Avg}$ the classical result for the average proper specific energy becomes

$$\varepsilon_e = \frac{E_e}{m_e} \cong \frac{1}{2} c (v_e)_{Avg} = \frac{1}{2} c^2 (v_e/c)_{Avg} \qquad (3.43)$$

3.6 CONCLUDING REMARKS REGARDING RELATIVISTIC FLIGHT

At this point in our development of relativistic flight mechanics and before we embark on space travel to far off destinations, it is useful to review some of the important assumptions and omissions associated mainly with the present chapter. In doing this, we will properly set the stage for the "fun" details of Chapter 4; expect that what follows may not answer all the questions the reader may have at this point but will suffice until we can properly deal with certain detailed subject matter best left for the next chapter.

Perhaps the most important assumption made so far is that our calculations are idealized as to such matters as efficiency and the issue of rocket mass versus fuel mass. No matter what we decide on for the propellant, it is necessary to assume that the rocket must be of massive proportions and that its mass must be predominantly that of the fuel necessary for relativistic flight, assuming it is rocket borne. In fact, it would be preferred that the fuel be, in some fashion, part of the rocket structure itself and that it be converted at very high efficiency to a directed propellant efflux. Without mentioning what mechanism is to be employed, we have favored a photonic propellant. This was a logical choice since it greatly simplified the rocket mechanics calculations. A photon propellant has the advantage of yielding the highest specific impulse (efflux velocity) possible, c, and it is constant independent of the velocity of its source. However, the downside is that such a propellant also yields the lowest thrust to efflux power ratio. Then to develop a sufficient thrust to accelerate a spacecraft to, say, $1g$ would require that the spacecraft be of enormous size and that the conversion efficiency of fuel to properly directed photonic energy be 100% or nearly so. This implies that the best fuel should involve a matter/antimatter reaction so as to convert mass to energy by $E = mc^2$ with near 100% efficiency. Obviously, the technology for such a conversion process has not yet been developed. But everything we know and use today has been technology dependent. So who is to say what the future will yield in the area of propulsion systems. Think of the technological developments that have occurred in the last 20, 50, or 100 years—it is remarkable! Chapter 6 is devoted to the exploration of alternative (actually exotic) propulsion systems that may soon find their use, perhaps by NASA, for example.

The profound conclusions asserted in this book are viewed as possible but only if technology of the future permits. Should this be so, however, we cannot rule out exploration of at least the near universe within the life-time of the astronauts. The assumptions made with regard to space flight in matter-free and gravity-free space along a straight trajectory are reasonable. We expect that the space flight will begin outside of our solar system beginning at $v = 0$ with respect to home earth and that the spacecraft will be accelerated to relativistic velocities. Furthermore, the assumption with regard to the astronaut's ability to measure certain parameters such as proper time of flight and proper acceleration is entirely plausible. Also, it is reasonable to assume that some data can be communicated by standard telemetric means between a moving spacecraft and the earth.

3.7 PROBLEMS

1. Given that the rocket equation is (3.9) and that $(m_{e0} - m_e)$ is the total fuel mass consumed during a given journey when m_e is the burnout mass at the end of that journey. If the normalized spacecraft velocity is $v_e/u_e = 2.5$, do the following: (a) calculate the burnout mass as a fraction of the total consumed fuel mass; (b) calculate the theoretical proper propulsive efficiency, η_P.
   Answers: (a) $m_e/(m_{e0} - m_e) = 0.0894$; (b) $\eta_P = 69.0\%$.

2. How much larger is the thrust to efflux power ratio $(F_e/P_e)_{Chem}$ for a chemical rocket than that for a photon rocket $(F_e/P_e)_{Photon}$ if the maximum efflux velocity for the chemical rocket is taken to be $u_e = 4000$ m s^{-1}?
   Answer: $(F_e/P_e)_{Chem} = 1.5 \times 10^5 (F_e/P_e)_{Photon}$.

3. A spaceship is moving through space at a proper velocity of $v_e = 1.83c$ as calculated by the astronauts in the spacecraft. (a) What velocity would be calculated for this spacecraft by the engineers on earth? (b) What is the instantaneous time dilation ratio dt/dt_e for this spacecraft?
   Answers: (a) $v = 0.950c$; (b) $dt/dt_e = 3.20$.

4. If a spacecraft is moving through space at a constant proper acceleration of $1g$, what will be the instantaneous acceleration, b, and velocity, v, as determined by earth at a proper Einstein number of 2.0?
   Answer: $b = 0.019$ m s^{-1}; $v = 0.964c$.

5. A spacecraft travels to Alpha Centauri, 4.3 light years distant from the earth, at an average proper Einstein number of 6.0. What is the total distance for this journey as calculated by the astronauts?
   Answer: 0.128 light years.

6. A trip is planned to a star known to be 100 light years distant from the earth. What must be the spacecraft's average proper velocity if the spacecraft is to reach the star calculated by the astronauts to be 5 light years distant from the earth?
   Answer: $5.37c$.

7. A spaceship on a journey to the star Vega, a distance of 26.5 light years from the earth. The spacecraft is moving at a constant proper velocity $v_e = 3.5c$. (a) Find the velocity, v, as determined from the earth. (b) At any point in the journey, find the ratio of the distance from spacecraft to earth to that from earth to spacecraft. (c) At any point in the journey, find

the ratio of distance from spacecraft to earth to that from spacecraft to star-Vega. (d) At any point in the journey, find the ratio of the distance from earth to Vega to that from spacecraft to Vega.

Answers: (a) $v = 0.998c$; (b) $dS_e^*/dS = 0.060$; (c) $dS_e^*/dS_e = 0.285$; (d) $dS/dS_e = 4.73$.

8. Calculate the proper specific energy required for a spacecraft to journey to star Vega given the information in Problem 7.

Answer: $\varepsilon_e = 4.79 \times 10^{10}$ GJ kg^{-1}.

CHAPTER 4
Space Travel and the Photon Rocket

In order to carry out the calculations required in the succeeding sections of this chapter, it will be helpful to the reader to cite the important equations taken from Chapters 2 and 3. To do this, we will simply list them in the order of occurrence with the same equation number previously used. Thus, if the reader wishes, he/she can easily locate a given equation in the corresponding chapter for further information as needed.

4.1 SUMMARY OF IMPORTANT EQUATIONS

Time dilation

$$dt = \frac{dt_e}{\sqrt{1 - v^2/c^2}}. \tag{2.11}$$

Distance contraction

$$dS_e^* = dS\sqrt{1 - v^2/c^2}. \tag{2.14}$$

Proper rocket velocity

$$v_e = b_e t_e \text{ (constant } b_e\text{)}. \tag{3.3}$$

Proper flight distance

$$S_e = b_e(t_e^2/2) = v_e^2/2b_e \text{ (constant } b_e\text{)}. \tag{3.4}$$

Proper mass ratio

$$\frac{m_{eo}}{m_e} = e^{v_e/c}. \tag{3.15a}$$

Proper photon energy

$$E_e = m_e c^2 = \hbar f = \hbar c/\lambda. \tag{3.16}$$

Proper photon thrust

$$F_e = \frac{dE_e/dt_e}{c} = \frac{P_e}{c}. \tag{3.18}$$

Photon specific impulse

$$I_{SP} = F_e/(dm_e/dt_e) = c = 3.00 \times 10^8 \text{ m s}^{-1}. \quad (3.19)$$

Photon rocket propulsive efficiency

$$\eta_{P, Photon} = \frac{v_e/c}{1 + v_e/c} \times 100. \quad (3.20)$$

Proper thrust/efflux power ratio

$$\frac{F_e}{P_e} = \frac{1}{c} = 3.33 \times 10^{-9} \text{ s m}^{-1}. \quad (3.22)$$

Einstein number

$$\frac{v}{c} = \left[\frac{1 - e^{-2v_e/c}}{1 + e^{-2v_e/c}}\right] = \tanh\left(\frac{v_e}{c}\right). \quad (3.32)$$

Elemental time of flight ratio

$$\frac{dt}{dt_e} = \cosh\left(\frac{v_e}{c}\right). \quad (3.33)$$

Acceleration ratio

$$\frac{b}{b_e} = \cosh^{-3}\left(\frac{v_e}{c}\right). \quad (3.35)$$

Elemental distance ratio

$$\frac{dS}{dS_e} = \frac{\sinh(v_e/c)}{v_e/c}. \quad (3.36)$$

Elemental distance, Einstein number, and velocity ratios

$$\frac{dS_e^*}{dS_e} = \frac{v/c}{v_e/c} = \frac{v}{v_e} = \frac{\tanh(v_e/c)}{v_e/c}. \quad (3.38)$$

Proper specific energy

$$\varepsilon_e = \frac{E_e}{m_e} = c^2(e^{(v_e/c)_{Avg}} - 1) \cdot \cosh(v_e/c)_{Avg}. \quad (3.41)$$

4.2   THE FLIGHT PLAN AND SIMPLIFYING ASSUMPTIONS

Here again, we find it advantageous to make some simplifying assumptions so as to simplify the calculations. Shown in Fig. 4.1 is a depiction of the flight plan from the earth to the destination star X as determined by the astronauts. It is assumed that the initial proper velocity

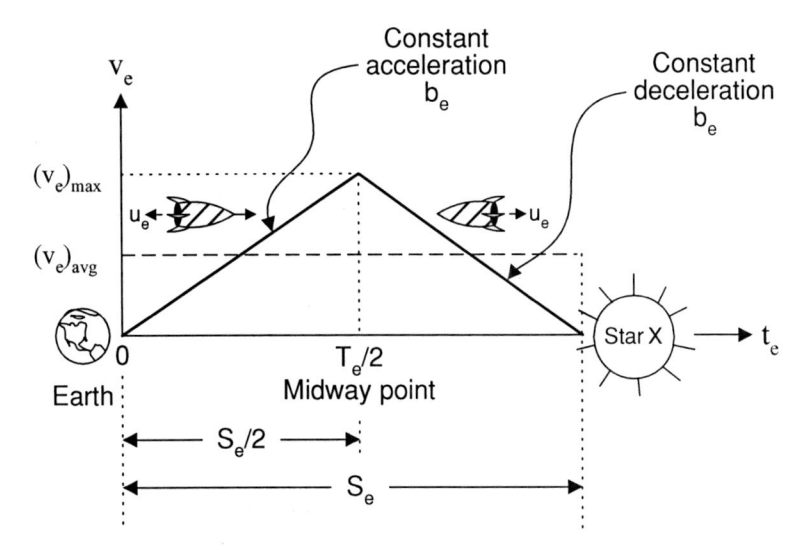

FIGURE 4.1: Flight plan for space vehicle showing a maximum proper velocity reached at midpoint between earth and star X and an average proper velocity for the entire journey to star X as determined by the astronauts

is known by the astronauts to be $v_{eo} = 0$ at $t_e = 0$ for the beginning of the flight. Also, we assume that the proper velocity, v_e, increases linearly with proper time t_e to produce a constant proper acceleration of $b_e = 10$ m s^{-2} or $1g$ in agreement with Eq. (3.3). At midpoint, $S_e/2$, in the journey as calculated by the astronauts by using Eq. (3.4), the proper velocity reaches its maximum value of $(v_e)_{\max}$ at a proper time of flight measured to be $T_e/2$. Thereafter, the spacecraft is flipped about and decelerates under constant proper deceleration $b_e = 10$ m s^{-2} to the star, at which point the proper velocity returns to zero. Since the proper velocity varies linearly with proper time (Fig. 4.1), the astronauts would know that their proper Einstein number is $(v_e/c)_{Avg} = (v_e/c)_{\max}/2$ over the course of their journey. On earth, similar calculations and measurements are made, but which are in no way equivalent in magnitude to those determined by the astronauts. Thus, the earth observations yield a midpoint distance to star X of $S/2$ at a time of flight $T/2$ and a maximum velocity of $(v)_{\max}$ determined at that point. The earth versus astronaut differences in Einstein numbers (normalized velocities), times of flight, accelerations, distances, and velocities are governed by Eqs. (3.32) through (3.38) and will be the subject of discussion in the sections that follow. The specific energy requirement for space flight to star X is given by Eq. (3.41).

The return-to-earth journey requires a reversal of the flight plan to star X just described. Thus, the total round trip would be over a proper distance of $2S_e$ at a corresponding total proper time of flight $2T_e$ as determined by the astronauts. The earth observers, on the other hand, would determine a total roundtrip distance of $2S$ over a total time of flight of $2T$ both of which would differ greatly from the measurements and calculations made by the astronauts.

There are other assumptions that are inherent in the flight plan just described and depicted in Fig. 4.1. We assume that the spacecraft is equipped with calibrated accelerometers and chronometers for the measurements of proper acceleration, b_e, and proper time of flight, t_e. Also, it is assumed that the astronauts have for their use the same standard measurement equipment as do the stay-at-home scientists on earth. It is further assumed that the spacecraft moves through gravity-free and resistance-free space at $1g$ acceleration along a straight path toward the destination star or galaxy. As stated previously in Section 3.6, we must finally assume that the space vehicle consists almost entirely of the fuel reactants (if rocket borne) and the means required to produce an appropriately directed photonic propulsive efflux converted at, or very nearly at, 100% efficiency from the reactants. The nature of the reactants and the means by which a photonic efflux might be produced are left for discussion in Chapter 6. In short, a massive photon rocket is required to move under ideal conditions to its destination according to the flight plan given in Fig. 4.1. Other exotic transport systems, those that require little or no reaction fuel mass and hence no propellant, are briefly discussed in Chapter 6. In all cases the flight calculations are subluminal within the framework of special relativity.

4.3 COMPARATIVE DISTANCES, FLIGHT TIMES AND ROCKET MASS RATIOS TO VARIOUS STAR/PLANET SYSTEMS, GALAXIES AND BEYOND—CALCULATION PLAN

Calculation plan. To begin the calculations, it is necessary to establish what is known. We know to fair accuracy the distance, S (in light years), from earth to various target destinations within and outside our own galaxy. Also, we set the proper spacecraft acceleration at $b_e = 10$ m s^{-2} or $1g$, for the comfort of the astronauts. We assume that the distances to target destinations within our galaxy remain relatively constant with time. However, distances to target destinations outside of our galaxy can only be roughly estimated since the universe is expanding at an increasing rate. Knowing b_e and the distance, S, from the earth to the various target destinations permits us to calculate $(v_e/c)_{Avg}$ from which we can calculate S_e, the respective times of travel T and T_e, and the initial-to-final proper mass ratios m_{e0}/m_e all as functions of $(v_e/c)_{Avg} = (1/2)(v_e/c)_{max}$, where $(v_e/c)_{max}$ is reached at midpoint in the journey. This is the *calculation plan* to be used hereafter.

We begin by using Eqs. (3.4) and (3.36) to establish the earth-to-destination distance, S, as a function of the average proper Einstein number, $(v_e/c)_{Avg}$. To do this, we first recognize that the constant acceleration b_e occurs only over the distance, $S_e/2$. Next, we normalize v_e in Eq. (3.4) by converting it to the average proper Einstein number $(v_e/c)_{Avg}$, which is valid over the proper distance S_e since the proper velocity v_e increases linearly with time to midpoint then decreases linearly with time to the destination star as in Fig. 4.1. Then, by introducing the integrated form of Eq. (3.36) together with the normalized form of Eq. (3.4) there results

$$S_e = \frac{(v_e/c)^2_{Avg}}{2b_e} \cdot c^2 = \frac{(v_e/c)_{Avg}}{\sinh(v_e/c)_{Avg}} \cdot S \qquad (4.1)$$

where we recall that the journey to the destination begins with $S_{e0} = S_0 = 0$ at time $T_{e0} = T_0 = 0$. Now, solving for S gives

$$S = \frac{c^2}{2b_e}(v_e/c)_{Avg} \sinh(v_e/c)_{Avg}, \qquad (4.2)$$

which yields

$$S = 4.50 \times 10^{15}(v_e/c)_{Avg} \sinh(v_e/c)_{Avg} \text{ in m}$$

or

$$S = 0.476(v_e/c)_{Avg} \sinh(v_e/c)_{Avg} \text{ in Lt Yr} \qquad (4.3)$$

Here we have taken $c = 3 \times 10^8$ m s^{-1} and $b_e = 10$ m s^{-2}, where 1 Lt Yr = 9.46×10^{15} m. Knowing S, Eq. (4.3) permits $(v_e/c)_{Avg}$ to be calculated, according to the calculation plan.

The proper distance, S_e is calculated directly from Eq. (4.1) as follows:

$$S_e = \frac{c^2}{2b_e}(v_e/c)^2_{Avg} = 0.476(v_e/c)^2_{Avg} \text{ in Lt Yr.} \qquad (4.4)$$

The calculation of travel times, T_e and T, as functions of $(v_e/c)_{Avg}$, is made in a somewhat similar manner to those of S_e and S. Beginning with Eq. (3.3) we write

$$T_e = \frac{v_e}{b_e} = \frac{c}{b_e} \cdot (v_e/c)_{Avg}. \qquad (4.5)$$

Then introducing $c = 3 \times 10^8$ m s^{-1} and $b_e = 10$ m s^{-2} there results

$$T_e = \frac{3.0 \times 10^8}{10}(v_e/c)_{Avg} = 3.0 \times 10^7(v_e/c)_{Avg} \text{ in s}$$

or

$$T_e = 0.951(v_e/c)_{Avg} \text{ in Yr} \qquad (4.6)$$

where 1 Yr = 3.154×10^7 s. The expression for T results by combining the integrated form of Eq. (3.33) with Eq. (4.6) to give

$$T = T_e \cosh(v_e/c)_{Avg} = 0.951(v_e/c)_{Avg} \cosh(v_e/c)_{Avg} \text{ in Yr.} \quad (4.7)$$

Finally, the initial-to-final propellant mass ratios are calculated from Eq. (3.15a) given by

$$\frac{m_{eo}}{m_e} = e^{(v_e/c)_{Avg}} \quad (4.8)$$

where again the average proper Einstein number is used.

4.3.1 Star/Planet Systems, Galaxies, and Beyond

As stated in Section 1.3, there are currently about 160 planets associated with nearly 100 stars/planetary systems that have been discovered over the past few years. These numbers are growing at an increasing rate with time. Shown in Table 4.1 are ten of these star/plant

TABLE 4.1: List of Star/Planetary Systems Giving the Associated Star, Its Distance from Earth, Its Magnitude, and the Estimated Mass of an Associated Planet Compared to Jupiter (= 1) Except Where Indicated Otherwise

STAR NAME	DISTANCE, FROM EARTH, S IN LIGHT YEARS (LT YR)	STAR MAGNITUDE	MASS OF ASSOCIATED PLANET (X JUPITER = 1)
Epsilon Eridani	10.4	3.73	0.1
Gliese 876	15	10.2	7.5 (x Earth = 1)
GJ 436	30	10.7	0.07
55 Cancri	44	5.95	18 (x Earth = 1)
Iota Draconis	100	3.3	8.7
HD 154857	222	7.25	1.8
HD 73526	323	9	3
GSC 02652-01324	500	12	0.75
OGLE-TR	5,000	16.6	0.9
OGLE-2005-BLG-071	15,000	?	≥3

systems chosen from a NASA/JPL list, which is updated frequently on the web (see Endnotes). These ten star/planet systems are listed in the order of increasing distance from the earth. The star magnitude gives the relative brightness of the star as viewed from the earth with the brightest stars having the lowest magnitudes. Star magnitudes equal to and greater than 6 normally cannot be seen with the unaided eye. The mass associated with a given planet is relative to that of Jupiter (= 1) unless otherwise given a mass rating relative to that of Earth (= 1).

To help understand the relative values as calculated from Eqs. (4.3), (4.4), (4.6), (4.7), and (4.8) as they pertain to the ten selected star/planetary systems given in Table 4.1, we provide additional information in Table 4.2. Here, the average proper Einstein numbers $(v_e/c)_{Avg}$ were calculated from Eq. (4.3) knowing the distance S of each star from the earth (see Redshift in Section 5.2). Then by using the values for $(v_e/c)_{Avg}$, calculations of S_e, T, T_e, and m_{e0}/m_e

TABLE 4.2: List of Star/Planetary Systems Giving the Star Name, Average Proper Einstein Number for Travel to the Star, The Respective Distances S and S_e to the Star as Measured from Earth and Calculated by the Astronauts, the Respective Times of Travel T and T_e to the Star as Measured from Earth and by the Astronauts, and the Initial-to-Final Mass Ratio for the Journey to the Star (The Values are Based on a Constant Proper Acceleration/Deceleration Equal in Magnitude to $b_e = 10 \text{ m s}^{-2}$)

STAR NAME	S (LT YR)	$(v_e/c)_{Avg}$	S_e (LT YR)	T (YR)	T_e (YR)	m_{e0}/m_e
Epsilon Eridani	10.4	2.76	3.63	20.8	2.63	15.8
Gliese 876	15	3.035	4.38	30.1	2.89	29.8
GJ 436	30	3.57	6.07	60.3	3.40	35.5
55 Cancri	44	3.87	7.13	88.3	3.68	47.9
Iota Draconis	100	4.53	9.77	200	4.31	92.8
HD 154857	222	5.19	12.8	443	4.94	179.5
HD 73526	323	5.51	14.4	647	5.24	247.1
GSC 02652-01324	500	5.88	16.5	1000	5.59	357.8
OGLE-TR	5,000	7.89	29.6	10,000	7.50	2,660
OGLE-2005-BLG-071	15,000	8.87	37.4	30,000	8.43	7,110

are found as presented in Table 4.2. Note that all calculations are based on a constant proper acceleration/deceleration equal in magnitude to $b_e = 10 \text{ m s}^{-2}$.

Note the contrast between distances and times of travel as determined from the earth with those determined by the astronauts. For example, the journey from the earth to Iota Draconis, a distance of 100 light years, as measured from the earth, is calculated by the astronauts to be a distance of only 9.77 light years when traveling at an average proper velocity of $4.53c$. The time elapsing for this one-way journey to the star at warp 4.53 is calculated to be 4.31 years astronaut time but about 200 years earth time. The round trip back to the point of origination requires that the data presented in Table 4.2 be doubled. So for a round trip to Iota Draconis, a distance of 200 light years measured from the earth is less than 20 light years as calculated by the astronauts. For this round trip, the astronauts would return "home" after 8.62 years only to find that the earth had undergone a time period of 400 years with very significant changes in the earth they once knew. The initial propellant mass, if rocket borne, will have to be at least 93 times the mass on arrival to the star and at least 186 times that on arrival back to Earth. Here, as indicated earlier, we assume that the mass of the spacecraft is essentially that of the propellant fuel, and that it is converted to propulsive power at nearly 100% efficiency, assuming ideal conditions. The journey just described originates just outside our solar system and ends at the star Iota Draconis. We assume that the distance from the earth to the starting point for the journey to be negligibly small compared to the distance to the star. Thus, the journey begins far enough outside of our solar system so as to be in gravity-free space yet close enough to Earth/solar system such that that distance can be considered negligible compared to the total journey distance to the star.

Shown in Fig. 4.2 are plots of Eqs. (4.3), (4.4), (4.6), (4.7), and (4.8) exceeding the range of values presented in Table 4.2. The method used to generate these plots is the same as the calculation plan given earlier. Log–log and semilog plots (some ×10) are used to accommodate the large magnitudes of distance traveled from the earth, time of travel measured from the earth, and the fuel mass ratio for the spacecraft. This permits the proper values of distance and time, as determined by the astronauts, to be placed on the same plot for comparison purposes. Two example trips are shown in Fig. 4.2, one to star Epsilon Eridani and the other to star OGLE -TR. Given the distances from the earth in light years, the intercepts of the various curves with the vertical "S" lines provide estimates of the remaining five quantities that can be compared with those in Table 4.2. To practice the method just described, the reader should apply the data given for the star Iota Draconis, which is 100 (10^2) light years from the earth. Thus, this vertical "S" line gives a value for $(v_e/c)_{Avg}$ of about 4.5, and provides the values for the remainder of the five quantities. As other interesting star/planet systems become known and are within the range depicted in Fig. 4.2, the various earth/astronaut data can be readily estimated from this figure. More "accurate" values must be taken directly from Eqs. (4.3), (4.4), (4.6), (4.7), and (4.8).

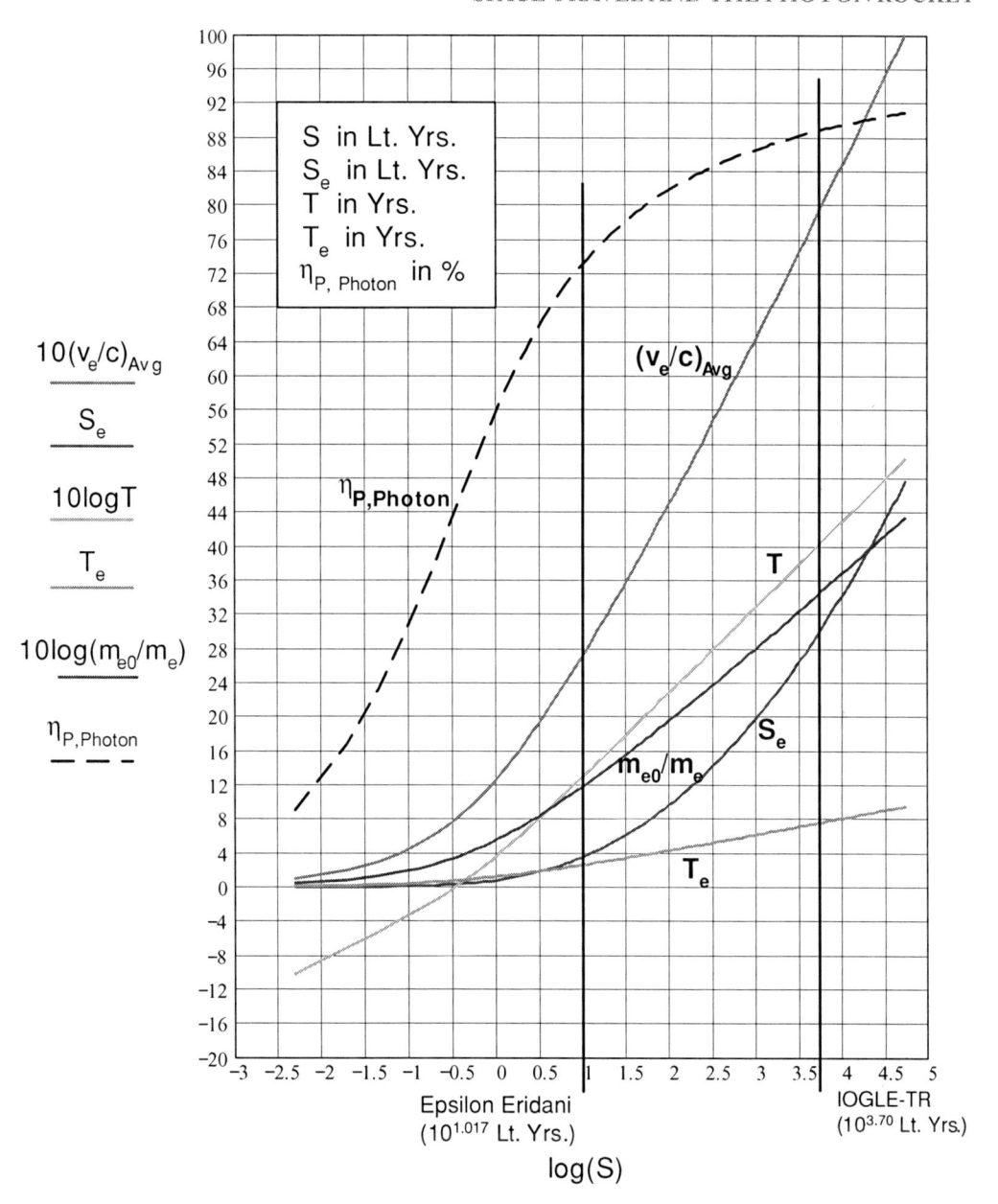

FIGURE 4.2: Plots of Eqs. (4.3), (4.4), (4.6)–(4.8) as they pertain to Table 4.2. The example trips to Epsilon Eridani and OGLE-TR are indicated by verticle black lines at the indicated distances in light years. The intercepts of the various curves with the vertical lines give the approxomate values for the remaining five quantities provided in Table 4.2

TABLE 4.3: List of Four Target Destinations Giving the Destination Name, Average Proper Einstein Number for Travel to the Destination, the Respective Distances S and S_e to the Destination as Measured from Earth and calculated by the Astronauts, the Respective Times of Travel T and T_e to the Destination as Measured from Earth and by the Astronauts, and the Initial-to-Final Propellant Fuel Mass Ratio for the Journey to the Destination (The Values are Based on a Constant Proper Acceleration/Deceleration Equal in Magnitude to $b_e = 10 \text{ m s}^{-1}$)

DESTINATION	S (LT YR)	$(v_e/c)_{Avg}$	S_e (LT YR)	T (YR)	T_e (YR)	m_{e0}/m_e
Alpha Centauri	4.3	2.14	2.18	8.77	2.03	8.50
Center of Our Galaxy	$30,000 \cong 10^{4.5}$	9.49	42.9	5.97×10^4	9.02	1.32×10^4
Messier 31 Galaxy	2.25×10^6	13.5	86.8	4.68×10^6	12.8	7.29×10^5
Circumnavigation of the Universe	5×10^{10}	22.9	250	9.6×10^{10}	21.8	8.8×10^9

Shown in Table 4.3 and Fig 4.3 are four other interesting destinations given in the order of increasing distance from Earth. Alpha Centauri (or Alpha Proxima) is included here since it is our nearest star but is not known at this time to have any planets associated with it. To reach Alpha Centauri, traveling at an average proper Einstein number of 2.14, would require about 2.0 years astronaut time but nearly 9 years earth time, and would require that the initial propellant fuel mass be nearly nine times that for burnout at arrival to the star. Also, from Fig. 3.5 the average velocity, as determined from the earth, would be about $0.97c \cong 180,000$ miles per second. The example trip to Alpha Centauri is featured in Fig. 4.3 from which the values of the quantities in Table 4.3 can be approximately verified.

As a second example, the flight to our nearest galaxy, the Messier 31, M31 galaxy (after Charles Messier, astronomer), is illustrated in Fig. 4.3. This great spiral galaxy lies in the constellation Andromeda and is sometimes referred to as the Andromeda galaxy or nebula. Its approximate distance from Earth is known to be about $S \cong 10^{6.35} \cong 2.25 \times 10^6$ Lt Yr = 690 kpc, where kpc is the abbreviation for kiloparsecs (see Appendix A). The vertical "S" line at $\log(S) = 6.35$, or $S = 10^{6.35} \cong 2.25 \times 10^6$ Lt Yr, intersects the various curves to give the approximate values for $(v_e/c)_{Avg} = 13.5$, S_e, T, T_e and m_{e0}/m_e. Recall that Eq. (4.3) is used to determine the average proper Einstein number knowing the distance S. So at an average proper spacecraft velocity of $(v_e)_{Avg} = 13.5c$ the astronauts would measure a proper flight time of

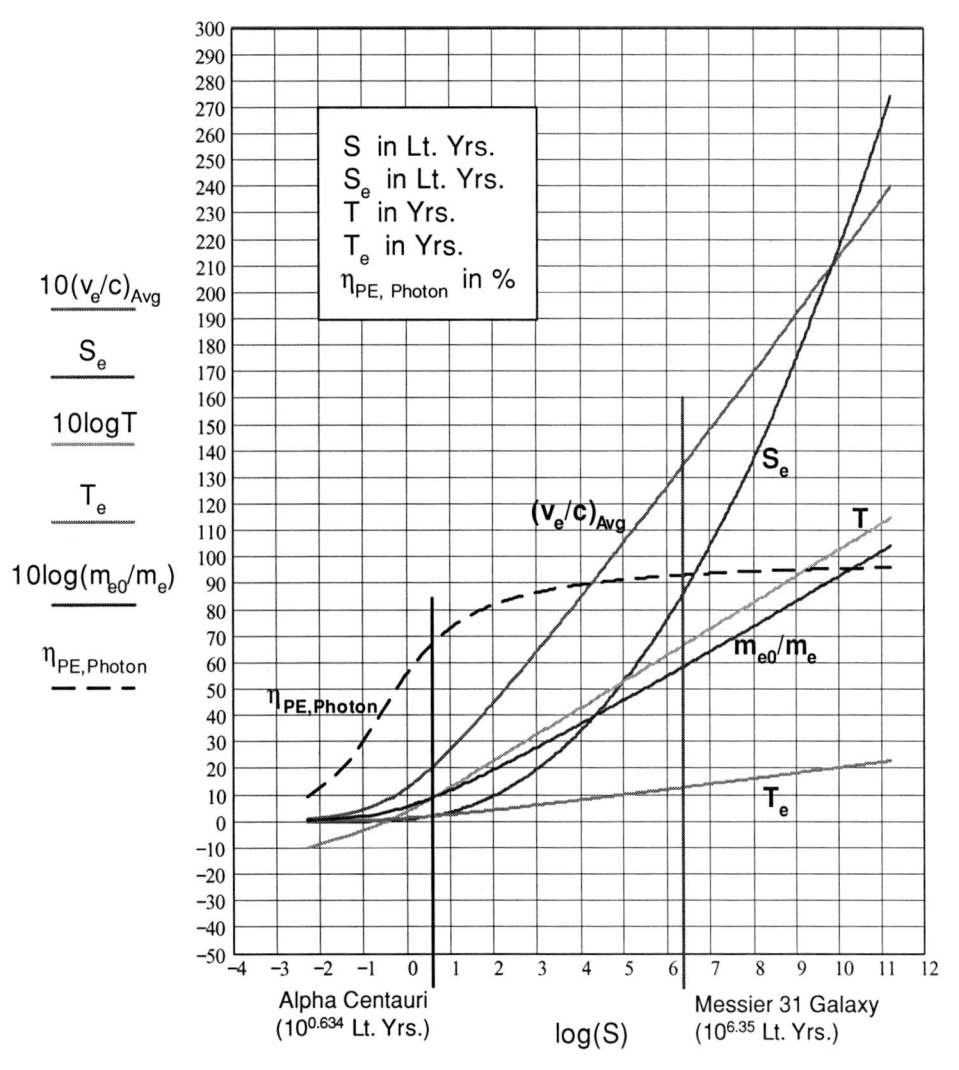

FIGURE 4.3: Plots of Eqs. (4.3), (4.4), (4.6)–(4.8) as they pertain to Table 4.3. The example trips to Alpha Centauri and to the M31 galaxy are indicated by verticle black lines at the indicated distances in light years. The intercepts of the various curves with these vertical lines give the approximate values for the remaining five quantities provided in Table 4.3

12.8 Yr (about 5 million years earth time!) over a calculated proper journey distance of about 86.8 Lt Yr.

As a third example, the distance from the earth to the center of our galaxy (the Milky Way) is known to have an upper value of about $10^{4.48} \cong 3.0 \times 10^4$ Lt Yr (in the direction of the constellation Sagittarius) from which the average proper Einstein number is found to be about

$(v_e/c)_{Avg} \cong 9.5$. But at $v_e/c \cong 9.5$ the astronauts calculate a proper distance of 42.9 light years with an estimated proper initial-to-final fuel mass ratio m_{eo}/m_e of about $10^{4.1}$ or 1.3×10^4. Thus, the initial proper mass must be about 10^4 times larger than the instantaneous burnout mass at the conclusion of the journey. The astronauts calculate that it will take about 9 proper years to complete the journey but are mindful that nearly 60,000 years will have passed by Earth time. The round trip back to Earth would require doubling the above numbers. Accordingly, the astronauts would return to Earth after about 18 years their time only to find that 120,000 years had passed Earth time—a frightening prospect to them, no doubt. The reader should use this example to further test his/her understanding in the use of Fig. 4.3.

There is much left to be discussed none the least of which are the practical considerations involved in carrying out the journeys described above. Most of this discussion will be left to Section 4.5. However, in the meanwhile we will discuss the consequences of particulate effluxes, the prospect of a varying speed of light, and the consequences of varying proper acceleration.

4.4 CONSEQUENCES OF PARTICULATE EFFLUX VELOCITIES, VARYING SPEEDS OF LIGHT, AND ACCELERATION

4.4.1 Consequences of Particulate Propulsion Systems

In the early development of relativistic rocket mechanics, the proper efflux velocity was expressed by symbol u_e without regard to the nature of the efflux. Such particulate propulsion systems include chemical, nuclear fission, and nuclear fusion rockets as examples. The initial-to-instantaneous proper fuel mass ratio, the specific impulse, propulsive efficiency, and the thrust to efflux power ratio were all cast in terms of u_e. Thus restating Eq. (3.9) or with reference to Eq. (C.8) in Appendix C, the rocket equation for the propulsion system of efflux velocity u_e is

$$v_e = -u_e \ln \frac{m_e}{m_{eo}} \qquad (3.9)$$

or

$$\frac{m_e}{m_{eo}} = e^{-v_e/u_e}. \qquad (4.9),(C.8)$$

When Eq. (4.9) is introduced into Eq. (3.31), the results show that Eq. (3.32) applies to particulate efflux propulsion as well as to photon efflux propulsion. In fact, Eqs. (3.32) through (3.38) all apply equally well to both particulate and photon propulsion systems under ideal conditions. This being true and excepting the flight plan in Fig. 4.1, it follows that Figs. 4.2 and 4.3 also apply equally well to both particulate and photon propulsion systems under the condition that $b_e = 10$ m s^{-2}. Given that the proper spaceship velocity, v_e, can exceed the efflux velocity, $v_e > u_e \ll c$ (see Fig. 3.2), it becomes theoretically possible for the spacecraft to achieve proper Einstein numbers greater than unity, that is, $v_e/u_e > 1$, assuming that u_e is high enough.

As an example, the journey to Alpha Centauri could take place in a particulate rocket propulsive system but with $u_e \ll c$ and vastly increased transit times among other limiting effects. But our calculations are based on the presence of an efflux from the plane of the spacecraft's nozzle independent of the propulsion system used, particulate or photonic.

So what are the fundamental differences between particulate and photon propulsion systems? Unlike photons, particles have a rest mass requiring that $u_e < c$. Therefore, as u_e approaches the speed of light (never to reach it), the time rate of change of linear momentum per particle is subject to relativistic dilation which increases the proper thrust F_e according to

$$F_e = \frac{d}{dt_e}\left(\frac{m_0}{\sqrt{1 - u_e^2/c^2}}\right) u_e, \qquad (4.10)$$

a result valid for constant u_e, where m_0 is the rest mass of the efflux particle. Thus, the proper thrust, F_e, increases rapidly with u_e as the velocity approaches the speed of light, c. However, at proper efflux velocities $u_e \ll c$, typical of chemical propellants, F_e increases nearly linearly with u_e, but the proper thrust to efflux power ratio falls off with u_e according to $F_e/P_e = 2/u_e$ given by Eq. (3.14). Then, as u_e approaches c, F_e/P_e again increases with u_e due mainly to the relativistic dilation of momentum as expressed in Eq. (4.10), but remember that v_e is greatly limited by $v_e > u_e \ll c$.

Other differences emerge between the particulate and photonic propulsion systems. Perhaps the most dramatic difference exists in the presentation of proper propulsive efficiencies. Whereas the proper propulsive efficiency for the classical particulate propulsion systems reaches 100% at $v_e/u_e = 1$ and drops off rapidly thereafter for $v_e/u_e > 1$, the photonic proper propulsive efficiency reaches 50% at $v_e/c = 1$ and then slowly approaches 100% as v_e/c becomes very large. This is best illustrated in Figs. 3.2 and 3.4, respectively. So a particulate propulsive system is not only limited by $v_e > u_e \ll c$ but is also limited by the rapid decline of the propulsive efficiency for proper Einstein values $v_e/u_e > 1$. Remember that Eq. (3.20), applicable to photonic propulsion, can never apply to particulate propulsion systems due to the differences in the residual proper efflux power lost in space; compare Eqs. (3.12) and (3.20).

4.4.2 Consequences of a Varying Speed of Light

We know that the speed of light (electromagnetic radiation) is less in materials than in free space. The more dense the medium, the more reduction in light velocity. In Section 2.1, the speed of light is shown to be equal to c/n in a liquid whose index of refraction is n, and a "drag" on the speed of light results when light is propagating in a fast flowing liquid. Given that the refractive index in water is about 1.33, then light propagates in still water at a velocity of about $0.75c$, but travels in glass at about $0.67c$. In an ultracold collection of atoms near absolute zero (called a Bose–Einstein condensate) light is slowed to a mere 17 m s^{-1} or about $5.7 \times 10^{-8}c$. However, the speed of light in mass media (in excess of c in a vacuum) has never been demonstrated experimentally. But the question remains: Is Einstein's Postulate II hard and fast or will space exploration and experimentation prove that it can be violated?

From what we know about light, given all the experiments that have been performed regarding its velocity, the velocity of light is precisely constant in a vacuum equal to $c = 3.00 \times 10^8$ m s^{-1}. Yet there is much we do not know about the propagation of light in interstellar and intergalactic space, perhaps also here on earth. In 2000 a team at the NEC Research Institute in Princeton, New Jersey reported that a pulse of light was propagated through a gas-filled chamber at 310 times the speed of light in a vacuum (Discovery, Vol. 26, No. 12 Dec. 2005, p. 24). No conformations or independent studies of this work are known by this author to have been reported. It seems likely that the pulse of light was propagated through a refractive medium such as cesium gas. This would cause a forward shift in the peak of the light pulse such that the peak would be detected before it was actually emitted. This could be interpreted as a violation of relativistic physics if, in fact, the observed effect was due only to shifts in the wavefunction of the light pulse.

Equally interesting is the contribution of a young theoretical physicist, named João Magueijo, who has strayed from the mainstream of thought by espousing a theory based on extremely dense cosmic strings that he believes to be vestiges of the "big bang" and that crisscross the universe. He supports the position that the velocity of light near these strings greatly exceeds the velocity of light in a vacuum (Discovery magazine Vol. 24, No. 4, 2003, pp. 34–41). However, Magueijo is not advocating exceeding the velocity of light established within the region of these cosmic strings, meaning that the laws of special relativity would not be violated. A paper by Andres Albrecht and João Magueijo titled "Time varying speed of light as a solution to cosmological puzzles" published in 1999 (see Endnotes) appears to be the basis on which Magueijo bases his assertions. It is fair to say that their work has not enjoyed popular acceptance among their fellow theoreticians. At least, Albrecht and Magueijo's work and that of the team at the NEC Research Institute add to the growing notion that there is much left to be learned about our physical universe. For example, current thinking is that dark matter exists and that dark energy is responsible for the increased expansion of the universe. But what is dark matter and energy and how might they affect interstellar space travel, if at all? How might light propagate through this dark matter that is thought to constitute as much as 90% of the mass in the universe?

According to modern cosmology, the universe is expanding at an increasing rate that might ultimately result in objects moving beyond a *cosmic horizon* where they would no longer be visible because their speed would exceed the speed of light. However, it is space itself not the objects in it that is expanding. Interestingly, the most distant galaxies are barely detectable at approximately 13.7 billion light years distant or 13.7×10^9 Lt Yr $\cong 4.2 \times 10^3$ Mpc. According to Hubble's law given by Eq. (5.14) in Chapter 5, the apparent velocity of recession of these galaxies would be about $v \cong 71 \cdot (4.2 \times 10^3) = 2.98 \times 10^5$ km s^{-1} = 2.98×10^8 m s^{-1} or about $0.99c$. Question: Is it possible that another universe extends beyond the so-called cosmic horizon where cosmic objects travel at velocities in excess of the speed of light as we know it in our universe?

So let us make the seemingly absurd assumption that the velocity of light, c (meaning electromagnetic radiation), will be found to be different in interstellar space than 3.00×10^8 m s^{-1} as measured on earth in a vacuum. What effect, if any, would this have on the idealized calculations given in Section 4.3? To demonstrate this, let us set the speed of light to a distant star equal to, less than, and greater than the velocity of light in a vacuum. In doing this, we will assume that the laws of special relativity remain valid. The object's velocity is not allowed to exceed a given value for the speed of light which is taken to be constant for each calculation. The distance, S, will remain the same for all calculations as will the flight plan given in Fig. 4.1 with $b_e = 10$ m s^{-2}. Table 4.4 lists the comparative results under different speeds of light for a spaceship journey to the star Iota Draconis known to be at 100 light years distance from the earth and known to have an associated planet approximately 8.7 times the mass of

TABLE 4.4: Comparative Results for a Trip to Star Iota Draconis Under Different Values for the Speed of Light, c' (All Values are Based on a Constant Proper Acceleration/Deceleration Equal in Magnitude to $b_e = 10$ m s^{-2} and Follow the Flight Plan Given in Fig. 4.1)

DESTINATION IOTA DRACONIS	S (LT YR)	$(v_e/c')_{Avg}$	S_e (LT YR)	T (YR)	T_e (YR)	m_{e0}/m_e
$c' = c$	100	4.53	9.77	200	4.31	92.8
$c' = c \div 2$	100	5.69	3.85	401	2.71	296
$c' = c \div 10$	100	8.50	0.344	1,986	0.81	4,910
$c' = c \times 2$	100	3.43	22.3	101	6.52	30.9
$c' = c \times 10$	100	1.27	76.8	23.2	12.1	3.56
$c' = c \times 100$	100	0.145	100	13.91	13.76	1.16
$c' = c \times 10^3$	100	0.015	100	14.27	14.27	1.015
$c' = c \times 10^4$	100	0.0015	100	14.27	14.27	1.002

Jupiter. The calculations are made by using the same calculation plan for each light speed as was used in Section 4.3.

An inspection of Table 4.4 indicates that an increase in the speed of light moves back the threshold for relativistic effects as determined by the astronauts, whereas a decrease in the speed of light has the reverse effect. The result is that an increase in the speed of light increases the proper values for time and distance, T_e and S_e, which finally reach terminal values as indicated. Note that the differences between the earth and proper values for time and distance ($T - T_e$ and $S - S_e$) are decreased with increasing light speed, but eventually reach a zero and constant difference, respectively. A decrease in the speed of light has the opposite effects. Other results of an increase in light speed include a reduction in the average proper Einstein number $(v_e/c')_{Avg}$ required under the flight plan of Fig. 4.1, and a decrease in the initial-to-final rocket (fuel) mass ration. Note that the mass ratio approaches unity (but never reaches it) due to the spacecraft's velocity that increasingly diverges from relativistic values, and that no account is taken of the payload and mass of the propulsion system engines and associated hardware. Overall, a significant increase in the speed of light is most favorable provided that the proper flight times, T_e, fall well within the lifespan of the astronauts. In the case of the 100 Lt Yr trip to Iota Draconis, the maximum time of travel ($T_e = T$) is but a little more than 14 earth years even if the velocity of light is taken to be much larger than $c' = c \times 10^4$. Thus, for an increase of light speed of $c' = c \times 10$ or greater and on return back to earth (doubling the values in Table 4.4), the astronauts need not be apprehensive about an inordinate time lapse on earth compared to theirs. Conversely, should the velocity of light be less than c, say for the case of $c' = c/2$, the astronauts would return to their earth which has aged at least by 800 years compared to their proper time elapse of about 5.4 years—a depressing thought that would seem not to inspire such a journey.

4.4.3 Consequences of a Varying Proper Acceleration

Unlike the velocity of light which we now take to be constant at $c = 3.00 \times 10^8$ m s^{-1}, the acceleration of the space vehicle can be changed purposefully within reasonable limits. The proper acceleration $b_e = 10$ m s^{-2} is chosen for the calculations in Section 4.3 because we are accustomed to the $1g$ force that has been exerted on our bodies over our lifespan. However, at this point in the technological development of space vehicle travel, we have no idea whether or not this acceleration can be sustained over the time periods required for interstellar travel. Reduction near weightlessness, $b_e \to 0$, would probably adversely affect the health of the astronauts over the proper flight times given in Table 4.3. Humans need near $1g$ force to maintain a healthy bone density and structure and to avoid serious infirmities. Conversely, a large increase in b_e might be impossible to sustain technologically speaking, and might also have adverse effects on the astronauts' health. However, current understanding of increased gravitational effects

TABLE 4.5: Comparative Results for a Trip to Star Iota Draconis Under Different Proper Vehicle Accelerations, b_e (The Flight Plan Given in Fig. 4.1 is Followed with the Velocity of Light, c, Held at 3.0×10^8 m s^{-1}, and the Distance from Earth to Iota Draconis, S, Held the Same at 100 Lt Yr)

DESTINATION IOTA DRACONIS	S (LT YR)	$(v_e/c)_{Avg}$	S_e (LT YR)	T (YR)	T_e (YR)	m_{e0}/m_e
$b_e = 2g = 20$m s^{-2}	100	5.10	6.19	199	2.43	164
$b_e = 1g = 10 ms^{-2}$	100	4.53	9.77	200	4.31	92.8
$b_e = 0.5\,g = 5$m s^{-2}	100	3.97	15.0	200	7.54	53.0
$b_e = 0.1\,g = 1$m s^{-2}	100	2.74	35.7	203	26.1	15.5
$b_e = 0.01\,g = 0.1$m s^{-2}	100	1.27	76.8	232	121	3.56
$b_e = 10^{-3}g = 0.01$m s^{-2}	100	0.451	96.8	473	429	1.57

subjected to humans over long periods of time is not well known. To add to our understanding of acceleration changes, we will next consider such changes as they relate to the idealized calculations in Section 4.3, but not to the health of the astronauts. In particular, we are interested in reduced values for b_e since they would likely be more easily sustained from a technological point of view.

Shown in Table 4.5 are the comparative results for the trip to Iota Draconis, a distance of 100 light years from the earth, under conditions of varying acceleration holding the speed of light constant at $c = 3.00 \times 10^8$ m s^{-1}. Accelerations between $g/2$ and $g/10$ appear to be optimal with respect to proper travel time and with respect to the initial-to-instantaneous mass ratio. However, a cursory inspection of the table indicates that decreasing the acceleration of the spacecraft below $0.1g$ moves the proper travel times for the astronauts out of range of a human life-time. Note that differences between earth and proper distances and between earth and proper travel times are decreased by a reduction in acceleration as are the proper Einstein numbers and the mass ratios.

For comparison purposes, consider a hypothetical particulate propulsive rocket that can achieve an average proper velocity of about $v_e \cong 1 \times 10^5$ ft s^{-1} or about 3×10^4 m s^{-1} (67,000 mi h^{-1}), given an efflux velocity of roughly the same order of magnitude. This is a classical case since $v_e \ll c$ and Eqs. (3.3) and (3.4) can be used to calculate the time of flight required. Accordingly, since $v_e/c \cong 3 \times 10^4 / 3 \times 10^8 = 1 \times 10^{-4}$, it would take the astronauts about 1,000,000 years for the 100 Lt Yr trip to Iota Draconis. Thus, it would take an increase in spacecraft velocity of

TABLE 4.6: Results for a Trip to Star Iota Draconis Under Constant Values for the Velocity of Light, $c' = c \times 10 = 30 \times 10^8$ m s^{-1}, and Proper Vehicle Acceleration $b_e = 0.1g = 1$ m s^{-1}, Following the Flight Plan Given in Fig. 4.1 with the Distance from Earth to Iota Draconis Given as $S = 100$ Lt Yr

DESTINATION IOTA DRACONIS	S (LT YR)	$(v_e/c')_{Avg}$	S_e (LT YR)	T (YR)	T_e (YR)	m_{e0}/m_e
$c' = c \times 10$ $b_e = 0.1g = 1$m s^{-2}	100	0.451	96.8	47.3	42.9	1.57

the order of 40,000 times that for the featured hypothetical rocket if the round-trip journey to Iota Draconis is to take place over the life-time of a human, say 50 Yr. Consequently, it is fair to say that conventional chemical rocketry, with or without multiple staging, can be ruled out as a practical means of interstellar travel.

4.4.4 Consequences of Combining Variable Light Speed with Variable Acceleration

It is interesting to combine the results of Tables 4.4 and 4.5. However, given the many possible combinations of speed of light and acceleration, we will give but one example. Shown in Table 4.6 are the typical travel parameters for a journey to Iota Draconis where the speed of light and vehicle acceleration are set at $c' = c \times 10 = 30.0 \times 10^8$ m s^{-1} and $b_e = 0.1g = 1$ m s^{-1}, respectively, and where the flight plan of Fig. 4.1 and calculation plan are followed. Note that the proper travel times are high but still within the life-times of the astronauts and, as expected, the difference between the travel times, as measured by the earth observers and the astronauts, approaches a common value. The low mass ratio reflects the low value of the proper Einstein number. Of course, there is no current justifiable reason to assume that the speed of light can be altered from the fixed value of c. But as discussed earlier in this section, we must take into account all possibilities whether or not they seem reasonable given the state of current knowledge of the universe and the laws of physics that we believe to be applicable. Furthermore, it is unknown if astronauts can endure an acceleration as low as $0.1g$ over many years of space travel.

4.5 "GIVE ME *WARP* THREE, SCOTTY"; PRACTICAL CONSIDERATIONS

In a Star Trek adventure it was a given that Captain Kirk could order Scotty to propel their spaceship at some warp speed and that the audience would accept this without questioning whether or not it made sense. Our task, to discuss some of the practical considerations regarding a technology that is yet to be discovered, is a daunting one and certainly open to questions; "a little knowledge can be dangerous." However, we will make an attempt to cite some of the practical problems based on assumptions made so far.

4.5.1 Efficiency Considerations

Admittedly, our treatment has been idealized so as to bring the important aspects of relativistic space flight to the forefront without engaging in unnecessary detail at that stage. For example, we have asserted that the fuel is converted to a propulsive efflux at or nearly at 100% internal efficiency. This, of course, is a huge oversimplification requiring that the generated particulates or photonic quanta be directionally collimated out of rocket's nozzle without losses of any kind including heat absorption by the reactor surroundings—an adiabatic process. We know that an adiabatic propulsion system is not possible with present technology though a near or quasiadiabatic conversion process may be feasible. We have also alluded to matter–antimatter reactants to generate a photonic efflux. What was not stated was that a matter–antimatter reaction emits electromagnetic radiation most likely over the entire spectrum (over 15 orders of magnitude in wavelength) including gamma radiation which is readily absorbed by the surrounding reaction chamber. In addition, muons, quarks, neutrinos, and other fundamental particles are likely to be emitted in a matter–antimatter reaction, and there is no known means of directionally collimating them. The result of nonadiabatic conversion together with particle mass losses is certain to reduce the total efficiency of producing a directionally collimated efflux given by

$$\eta_T = \eta_I \cdot \eta_P, \qquad (4.11)$$

where η_P is the propulsive efficiency applicable to photon or particulate propulsion as previously expressed by Eqs. (3.12) and (3.21). Recall that the internal efficiency, η_I, lumps together all the efficiency factors (thrust efficiency, thermal efficiency, frozen flow efficiency, etc.) that influence the production of a directed photon efflux in the rocket's engine or reaction chamber. However, only η_P can be quantified given the limited information available. For a photonic efflux, η_P is given by Eq. (3.20) and Fig. 3.4 as a function of the proper Einstein number. Over the range of v_e/c (about 2 to 23) given in Tables 4.2 and 4.3, η_P varies from 67% to 96% which set the upper limits for η_T. This, in turn, will affect the results given in Sections 4.3, 4.4, and 4.5. See Appendix C.2 for the effect of efficiency on the rocket equations.

4.5.2 Payload, Engine, Crew and Reactor Hardware Mass Considerations

The values given for the mass ratio m_{e0}/m_e result from our assumption that the propellant fuel mass must be wholly rocket borne and, given our present understanding of interstellar space, this may be a good assumption. This, in turn, requires that the rocket be of massive proportions leaving little room for payload, engine, and reactor hardware, etc. Let us assume that the reaction fuel mass is represented by M, and that the payload plus the propulsion engine, reactor hardware, fuel containers, and crew all are represented by P. Under these assumptions, the rocket equation (3.15), applicable to relativistic velocities, can be written as

$$\frac{v_e}{c} = \ln\left(\frac{M+P}{P}\right) \tag{4.12}$$

where $m_{eo} = M + P$ and $m_e = P$ is the burnout mass at the end of the journey. An example will help to better understand the rocket equation and its application. From Table 4.5, we take $v_e/c = 4.53$ for the spaceflight to Iota Draconis at 1g acceleration. Then by Eq. (4.12), there results

$$\left(\frac{M+P}{P}\right) = e^{4.53} = 92.8.$$

Therefore,

$$P = M(e^{v_e/c} - 1)^{-1} = M/(92.8 - 1) = M/91.8 = 0.011M.$$

This means that P can be about 1.1% of M giving a mass ratio of about 92.8, in agreement with the results for a 1g acceleration in Table 4.5. Of course, this assumes that the journey terminates with burnout at Iota Draconis. If it happens that P should equal or exceed the initial reaction fuel mass, a likely possibility, travel to stars beyond our nearest star Alpha Centauri would seem unlikely. But then again, future technology may come to our rescue in this regard. The real interest may focus on the prospect of using fuel available in space so as to reduce if not eliminate rocket-borne fuel. But that is an issue to be discussed in Chapter 6.

4.5.3 Energy Requirements

So far nothing has been said about the energy requirements for interstellar flights. Presented in Table 4.7 are the comparative values of the mass ratio m_{e0}/m_e and specific energy ε as calculated by using Eqs. (3.15a) and (3.42), respectively, for a journey to Alpha Centauri and the ten star/planet systems of Table 4.2 where the flight plan of Fig. 4.1 is followed at 1g acceleration/deceleration. Note that the mass ratio and the specific energy are both a function only of the average proper Einstein number, $(v_e/c)_{Avg}$. As a reminder and for the convenience of the reader, the mass ratio and proper specific energy equations are given as follows,

$$\frac{m_{eo}}{m_e} = e^{(v_e/c)_{Avg}} \tag{4.13}$$

and

$$\varepsilon_e = 9.0 \times 10^7 (e^{(v_e/c)_{Avg}} - 1) \cdot \cosh(v_e/c)_{Avg}. \tag{4.14}$$

TABLE 4.7: Comparative Values of Mass Ratio m_{e0}/m_e and Energy Density ε as Calculated by Using Eqs. (4.13) and (4.14), Respectively, for Alpha Centauri and the Ten Star/Planet Systems of Table 4.2

STAR NAME	S (LT YR)	$(v_e/c)_{Avg}$	m_{e0}/m_e	ε_e (GJ KG^{-1})
Alpha Centauri	4.3	2.14	8.5	2.91×10^9
Epsilon Eridani	10.4	2.76	15.8	1.06×10^{10}
Gliese 876	15	3.035	29.8	1.86×10^{10}
GJ 436	30	3.57	35.5	5.52×10^{10}
55 Cancri	44	3.87	47.9	1.01×10^{11}
Iota Draconis	100	4.53	92.8	3.83×10^{11}
HD 154857	222	5.19	179.5	1.44×10^{12}
HD 73526	323	5.51	247.1	2.74×10^{12}
GSC 02652-01324	500	5.88	357.8	5.75×10^{12}
OGLE-TR	5,000	7.89	2,660	3.21×10^{14}
OGLE-2005-BLG-071	15,000	8.87	7,110	2.28×10^{15}

As an example, consider a spacecraft voyage to Alpha Centauri at a distance of 4.3 light years from the earth that would follow the flight plan in Fig. 4.1. The spaceship will travel at an average proper velocity of $(v_e)_{Avg} = 2.14c$ and an average earth-calculated velocity of $(v)_{Avg} = 0.973\,c$ according to Eq. (3.32). Let us assume that the spaceship has a total initial mass of, say, 2×10^6 kg. Then by Eq. (4.12) the burnout mass would be roughly $P = (M+P)/e^{(v_e/c)_{Avg}} = 2 \times 10^6/8.5 = 0.235 \times 10^6$ kg or about 13.3% of the initial fuel mass, M. Therefore, for the trip, the average total mass of the spacecraft would be about 1.1×10^6 kg. Introducing this value into Eq. (3.42) gives for the total energy required

$$E_{Total} = m\varepsilon_e = (1.1 \times 10^6) \cdot 9 \times 10^7 (e^{(2.14)_{Avg}} - 1) \cdot \cosh(2.14)_{Avg}$$
$$= (1.1 \times 10^6) \cdot 2.91 \times 10^9 \cong 3.2 \times 10^{15}\,GJ.$$

To put this in perspective, let us assume that the entire earth generates about a terawatt of power which is 10^{12} J s^{-1} = 10^3 GJ s^{-1}. In one year this amounts to $10^3 \times 365 \times 24 \times 3600 \cong$

80 RELATIVISTIC FLIGHT MECHANICS AND SPACE TRAVEL

3×10^{10} GJ of energy generation. Thus, to propel a spaceship of 10^6 kg average mass to Alpha Centauri at an average earth-calculated speed of $0.973c$ would require the expenditure of 10^5 times more energy than is generated by the entire earth in one year. A voyage to Iota Draconis, at a distance of 100 light years, would require about 100 times more energy than that for a journey to Alpha Centauri, according to Table 4.7. Thus, these are indeed enormous energy requirements, but not impossible. Possibly, spacecrafts of the future may somehow be able to make use of the dark energy that is believed to exist in space thereby reducing the need for onboard power generation. A few exotic transport or propulsion systems that might dramatically change the nature and extent of power generation in future space vehicles are discussed in Chapter 6.

4.5.4 Life-Preserving and Health Issues of the Astronauts

In our abbreviated coverage of practical considerations, we must discuss the life-preserving issues of the astronauts that are likely to be encountered during extensive transit times in the closed system of a space vehicle. The information gained from astronauts in a space station orbiting the earth under weightless conditions is of little or no use for journeys of the type represented in Table 4.4. A manned trip to Mars would be helpful in this regard but would still be lacking in information and experience needed for an interstellar flight. Food, air, and water required to sustain life on interstellar flights lasting several years top the list of essentials that must be considered. Food would have to be effectively grown. Use of animals for food may pose a problem since the animals would be placed in competition with humans for food, air, and water. Plants need soil, fertilizer, carbon dioxide, and light for growth. The soil can be fertilized by using human excrement, light can be supplied by various power sources, and CO_2, conveniently exhaled by humans, can also be generated by chemical means. Breathable air requires about 20% oxygen which can be chemically separated from CO_2. Water is needed for both human consumption and plant growth. Recycling processes can be used to obtain water from urine and perspiration but this offers only a limited solution to the problem. Oxygen, obtained from CO_2, can be combined with H_2 in fuel cells to produce electrical power with a pure water byproduct. Water can also be obtained from the leaves, stems, and roots of some plants. All of the above will require power which must come from fuel used for the propulsion system. Thus, to sum up, nothing must be left to waist or be open ended in a closed system such as an interstellar spacecraft—everything must be interrelated such as to produce a viable closed-loop system of processes.

Proper shielding from radiation and from high-velocity particle bombardment is a problem we are not yet fully prepared to deal with. The radiation from a matter/antimatter, fusion or fission reactor may consist of heat, gamma radiation, X-radiation and from fundamental particles such as neutrinos, muons, and quarks. All can create a considerable danger to the astronauts, in particular gamma radiation. However, shielding from interstellar particle bombardment may be the most challenging. Consider that a dust particle impinging on a spacecraft at relativistic velocity could cause serious damage on and inside the space vehicle.

Finally, there is the issue of physical and mental health of the astronauts during a long-term journey to a distant star. The space vehicle must be large enough to accommodate the essential exercise and recreational facilities. Also, it must have the appropriate medical personnel and equipment to deal with a variety of unexpected illnesses, diseases, and accidents. Certainly, trained personnel should be present who can effectively deal with interpersonal conflicts and individual psychoses. Conjugal relationships between people of opposite gender should not be overlooked and spiritual needs should be met.

To minimize the number of personnel required for a given interstellar journey, it will likely be necessary to double up on the qualifications of some individuals. As an example, some of the technical specialists that are required may have to double as medical personnel or as psychological councilors. Critical personnel, those essential to the operation of the space vehicle, must have one or more backups in case of accidental death, injury, or functional debilitation.

4.6 PROBLEMS

1. A starship initially weighing 10^7 kg is to journey to the star/planet system Epsilon Eridani which is 10.4 light years distant from the earth. (a) Use the information in Table 4.7 to determine the initial fuel mass at the beginning of the journey, and the burnout mass as a percentage of initial fuel mass at journey's end. (b) Find the average total mass for the journey and the total energy required for the journey. Follow the flight plan of Fig. 4.1 with an acceleration/deceleration of $1g$.
Answers: (a) $M = 9.37 \times 10^6$ kg; $P = 6.80\%$ of M; (b) $m_{Avg} = 5.3 \times 10^6$ kg; $E = 5.6 \times 10^{16}$ GJ.

2. Suppose that a fly-by trip is planned to Iota Draconis 100 light years distant from the earth. To do this, it is desirable to maintain an acceleration of $b_e = 10$ m s^{-1} for the entire journey to the star. Given this information, calculate the following assuming ideal conditions: $(v_e/c)_{Avg}$; S_e (in Lt Yr); T (in Yr); T_e (in Yr); m_{eo}/m_e; and ε_e (specific energy in GJ kg^{-1}).
Answers: $(v_e/c)_{Avg} = 5.1$; $S_e = 6.19$; $T = 199$; $T_e = 2.42$; $m_{eo}/m_e = 82.0$; $\varepsilon_e = 6.0 \times 10^{11}$.

3. It is desirable for a spacecraft to travel to Alpha Centauri at a distance of 4.3 light years from the earth following the flight plan of Fig. 4.1, but at a constant acceleration of $b_e = g/10$. Recalculate the parameters given in Table 4.3 for comparison.
Answers: $S = 4.3$; $(v_e/c)_{Avg} = 0.891$; $S_e = 3.78$; $T = 12.1$; $T_e = 8.5$; $m_{eo}/m_e = 2.44$.

4. For a photon-propelled spacecraft journey to Iota Draconis, consider the effect that efficiency has on the trip parameters. To do this, assume that the total efficiency, given in Eq. (3.21), is $\eta_T = 75\%$ and that it will affect the initial spacecraft mass ratio which is $m_{e0}/m_e = 100$ assumed to be mostly fuel mass. Given this information and assuming ideal conditions, do the following: (a) calculate the photonic propulsive efficiency, η_P, and the internal efficiency, η_I; (b) recalculate the parameters in Table 4.2 as they are affected by total efficiency via the effect on the initial mass ratio of the spacecraft. Explain the value obtained for S.
Answers: (a) $\eta_P = 83.0\%$, $\eta_I = 90.4\%$; (b) $S = 155$; $(v_e/c)_{Avg} = 4.89$; $S_e = 11.4$; $T = 310$; $T_e = 4.65$; $m_{e0}/m_e = 133.3$; S is a pseudodistance resulting from the incremented mass ratio.

5. A spacecraft is to journey to the star Vega in the constellation Lyre 26.5 light years distant from the earth following a flight plan similar to Fig. 4.1 but for $b_e = b_e(t_e)$. As the journey proceeds, assume that the consumed fuel mass makes it possible for the spacecraft to move at an increasing proper velocity given by $v_e = at_e^2$ instead of $v_e = b_e t_e$ as in Fig. 4.1. Fit $v_e = at_e^2$ to the same limits $[(v_e)_{max}, T_e/2]$ as in Fig. 4.1 that yield $b_e = 10$m s^{-2}, set $a = 1$ m s^{-3}, and assign arbitrary values to the limits to make the calculations simple. Now, do the following: (a) Calculate the mean value $(v_e)_{Avg}$ of $v_e = at_e^2$ in terms of $(v_e)_{max}$. (b) Find the value for b'_e at $(v_e)_{Avg}$. (Hint: find the slope of the curve $v_e = at_e^2$ at $(v_e)_{Avg}$.) (c) Find the value for $(b'_e)_{max}$ at $(v_e)_{max}$. (d) Use the value for b'_e to evaluate the parameters in Table 4.2. (e) Find the value for $(v_e/c)_{max}$.
Answers: (a) $(v_e)_{max}/3$; (b) $b'_e = 11.55$m s^{-2}; (c) $(b'_e)_{max} = 20$ m s^{-2}; (d) $S = 26.5$; $(v'_e/c)_{Avg} = 3.58$; $S_e = 5.28$; $T = 53.0$; $T_e = 2.95$; $m_{e0}/m_e = 35.9$; (e) $(v_e/c)_{max} = 10.74$.

CHAPTER 5

Minkowski Diagrams, K-Calculus, and Relativistic Effects

5.1 MINKOWSKI DIAGRAMS

Up to this point, we have developed the special relativity principles from strictly a theoretical point of view. Now we introduce a geometrical interpretation based on the earlier work of the German mathematician Herman Minkowski who, like Einstein, concluded that time is a fourth coordinate in the spacetime continuum in which time and space are inherently related. Shown in Fig. 5.1 is a two-dimensional (2D) Minkowski diagram for time/one-dimensional (1D) space. The line OT represents the time coordinate as measured by the observer at rest, and the line OX is the x-axis space coordinate along which an event takes place. Neither line OY nor OZ (for a three-dimensional (3D) representation) is included so as to focus attention on the basic concepts involved in this geometrical interpretation of special relativity without unnecessary complications. Each point in the diagram represents an event that takes place at some point in time along OT and at some point in space along the event line, OX. Future events take place above event line OX while past events take place below line, OX. Thus, the big bang can be represented as an upside-down pear shape below the event line, OX.

Consider that we are an observer at rest along OT and that we are sending out light signals at regular intervals, $N_1, N_2, N_3, \ldots$, as shown in Fig. 5.1. Let these light signals be received at intervals N'_1, N'_2, N'_3, etc. by an observer in the moving rocket along the rocket line OR, as indicated in Fig. 5.1. So that all coordinates have the same units of length, it is useful to set the OT coordinate in units of $i\tau = ict$, an imaginary axis since $i = \sqrt{-1}$, where $\tau = ct$. This is done so as to avoid the possibility of turning time into a physical entity. Thus, in a generalized Minkowski diagram, we define spacetime events in the four-dimensional (4D) coordinate system x, y, z, ict so that the Pythagorean sum of squares is given by the invariant

$$s^2 = x^2 + y^2 + z^2 + (i\tau)^2 = x^2 + y^2 + z^2 + (ict)^2$$

or in 2D spacetime

$$s^2 = x^2 + (i\tau)^2 = x^2 + (ict)^2. \qquad (5.1)$$

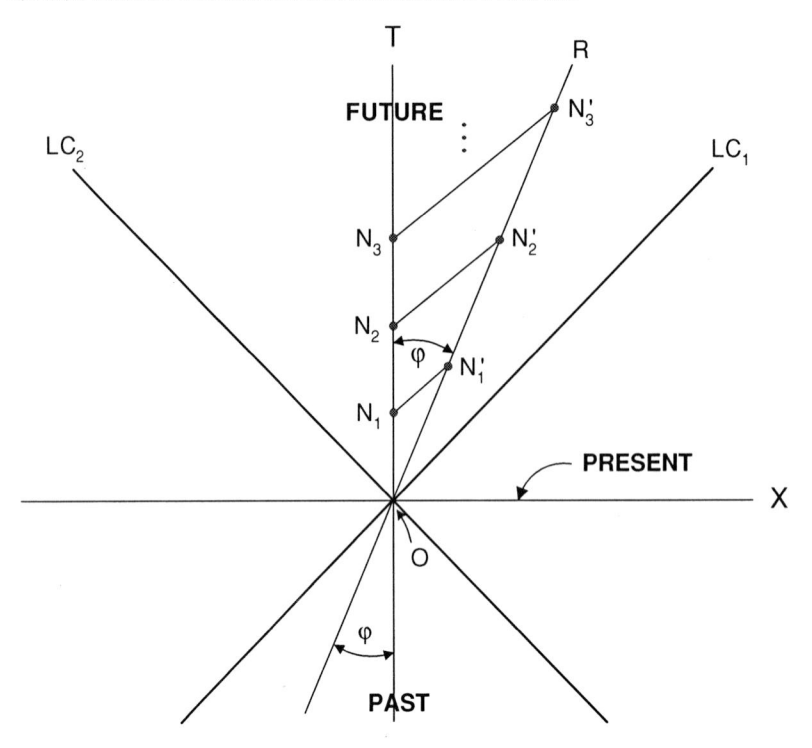

84 RELATIVISTIC FLIGHT MECHANICS AND SPACE TRAVEL

FIGURE 5.1: Minkowski spacetime diagram in 2D showing time line of stationary observer along OT, event line along OX (*x*-axis), moving rocket line OR, light cone extremities LC$_1$ and LC$_2$ in 2D, and light pulses $N_1, N_2, N_3, \ldots$ sent from stationary observer to rocket at intrevals N'_1, N'_2 and N'_3, etc.

In Fig. 5.1, the *light cone* is defined in 2D by $x = \pm i\tau$ such that LC$_1$ and LC$_2$ are the light cone extremities each at 45° relative to the OT and OX axes. This corresponds to setting $s^2 = 0$ in Eq. (5.1). Propagation of light rays to or through a given event must be confined within the light cone. When using imaginary time units, $i\tau = ict$, it is helpful to set the velocity of light to unity ($c = 1$) retaining for ct the same units used for the space coordinates.

Now, let $\Delta\tau_0$ be the time interval between successive pulses N_i sent from the stationary observer on OT to the moving rocket observer on OR, and let $\Delta\tau$ be the time interval between reception of the successive pulses N'_i. At relativistic velocities of the rocket, we can define the ratio of these two time events as the *K*-factor (sometimes Bondi *K*-factor, after Hermann Bondi) given by

$$K = \frac{\Delta\tau}{\Delta\tau_0}, \qquad (5.2)$$

where $\Delta\tau$ is necessarily greater than $\Delta\tau_0$. The observer in the rocket, traveling at a relative velocity, v, has a trajectory with respect to OX of

$$x = vt = \frac{v}{c}\tau \tag{5.3}$$

and an inclination relative to OT given by

$$\tan\varphi = \frac{vt}{ct} = \frac{v}{c}. \tag{5.4}$$

Note that by following Eq. (3.32) and including Eq. (5.4) the proper Einstein number becomes

$$\frac{v_e}{c} = \tanh^{-1}\left(\frac{v}{c}\right) = \ell n\sqrt{\frac{1+v/c}{1-v/c}} = \ell n\sqrt{\frac{1+\tan\varphi}{1-\tan\varphi}}, \tag{5.5}$$

which appears to suggest that the proper Einstein number can be regarded as a pseudorotation in the spacetime continuum. A rocket observer traveling at a different speed relative to a reference observer has a different point of view regarding which sets of points are simultaneous or which are at the same point in space—the time and space coordinates are different.

5.2 K-CALCULUS AND RELATIVISTIC EFFECTS AND MEASUREMENTS

With the aid of a Minkowski diagram, we will now use geometry to derive the time dilation expression given by Eq. (3.11), the relativistic Doppler effect, the redshift, and the addition of velocities given by Eq. (3.23) and generalized by Eq. (3.26). To accomplish these four tasks, we will follow our practice of using simple thought experiments, but now based on simple geometry.

5.2.1 Time Dilation

Consider the Minkowski diagram depicted in Fig. 5.2. Here, an observer at rest along time line OT exchanges light pulses with an observer in the rocket traveling at a velocity, v, on line OR. Let us assume that the two observers were initially together at origin O such that $\tau = \tau_0 = 0$. Thus, the stationary observer on OT sends a light pulse N at a time $\Delta\tau_0 = \tau_0$. The light pulse is received by the observer in the moving rocket at N' on OR at a time

$$\tau = K\tau_0. \tag{5.6}$$

The light pulse is immediately re-emitted by mirror reflection back to the stationary observer at N'' on OT at a time

$$\tau_1 = K\tau = K^2\tau_0. \tag{5.7}$$

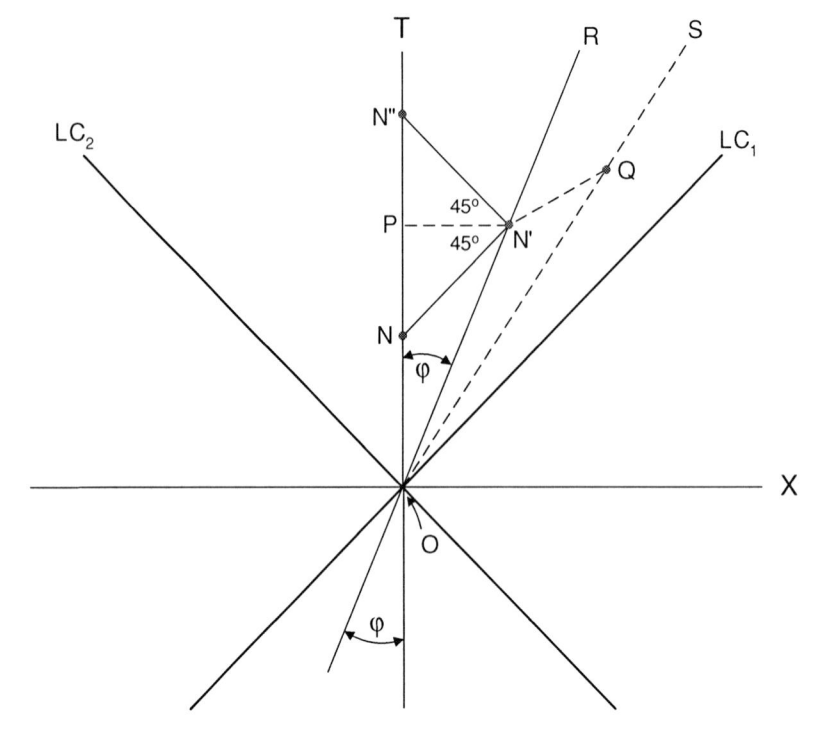

86 RELATIVISTIC FLIGHT MECHANICS AND SPACE TRAVEL

FIGURE 5.2: Minkowski spacetime diagram in 2D showing a pulse emitted by the stationary observer (N) on OT, its receipt by the moving rocket observer (N') on OR traveling at a velocity v with respect to the stationary observer, the reflection of that light pulse back to the stationary observer (N'') on OT, and the presence of a third observer on OS traveling at a velocity u with respect to the rocket observer on OR

Our first goal is to determine the K-factor from which the time dilation expression is obtained. From Fig. 5.2 and a little geometry we easily deduce the following:

$$ON = \Delta\tau_0 = \tau_0$$
$$NN' = N'N''$$
$$PN' = NP = PN'' = \frac{NN''}{2}$$
$$PN' = (OP)\tan\varphi = (OP)\frac{v}{c} = \frac{NN''}{2}$$
$$OP = ON + NP = ON + \frac{NN''}{2}$$
$$NN'' = \tau_1 - \tau_0 = K^2\tau_0 - \tau_0 = \tau_0(K^2 - 1)$$
$$(OP)\frac{v}{c} = \frac{NN''}{2} = \frac{\tau_0(K^2 - 1)}{2}$$

$$OP = \tau_0 + \frac{\tau_0(K^2 - 1)}{2} = \frac{\tau_0(K^2 + 1)}{2}$$

$$\frac{v}{c} = \frac{NN''/2}{OP} = \frac{K^2 - 1}{K^2 + 1}.$$

Or finally, by using Eqs. (5.5),

$$K = \sqrt{\frac{1 + v/c}{1 - v/c}} = e^{v_e/c}, \qquad (5.8)$$

which gives the K-factor in terms of both the Einstein number and the proper Einstein number.

Now since PN' is perpendicular to the time axis OT, the observer at rest will conclude that time event ON' is simultaneous with OP in real time, Δt. By using Eq. (5.6) there results

$$\Delta t = OP = \frac{\tau_0(K^2 + 1)}{2} = \frac{(K^2 + 1)}{2} \cdot \frac{\tau}{K}.$$

Therefore,

$$\frac{\Delta t}{\tau} = \frac{(K^2 + 1)}{2} \cdot \frac{1}{K} = \left(\frac{1 + v/c}{1 - v/c} + 1\right) \cdot \frac{1}{2K} = \frac{1}{2}\left(\frac{2}{1 - v/c}\right)\sqrt{\frac{1 - v/c}{1 + v/c}}$$

$$= \left(\frac{1}{\sqrt{1 - v/c}}\right)\left(\frac{1}{\sqrt{1 + v/c}}\right) = \frac{1}{\sqrt{1 - v^2/c^2}}.$$

Or finally,

$$\Delta t = \frac{\Delta t_e}{\sqrt{1 - v^2/c^2}} \quad \text{where} \quad \tau = \Delta t_e. \qquad (5.9)$$

Thus, by using the Minkowski diagram in Fig. 5.2, we have obtained by geometric means the time dilation effect given by Eq. (2.11).

5.2.2 Relativistic Doppler Effect

Note that Eq. (5.8) gives the relativistic Doppler shift for light and zero angle of viewing if $K = f_S/f_O$, where f_S is the frequency of light emitted by the moving source with velocity v relative to an observer, and f_O is the frequency of light as perceived by the observer. Thus, for zero angle of viewing, we can write

$$K = \frac{f_S}{f_O} = \sqrt{\frac{1 + v/c}{1 - v/c}} = \sqrt{\frac{c + v}{c - v}} = \frac{\sqrt{1 - v^2/c^2}}{1 - v/c}.$$

Solving for the frequency perceived by the observer there results

$$f_O = f_S\sqrt{\frac{1-v/c}{1+v/c}} = f_S\left(\frac{1-(v/c)}{\sqrt{1-v^2/c^2}}\right) \quad (5.10)$$

where Eqs. (3.1) and (3.16), given by $\lambda = c/f$ and $E = \hbar f$, permit Eq. (5.10) to be recast in terms of the wavelength, λ, and energy, E, respectively. Equation (5.10) applies to a source receding at a velocity, v, relative to a moving (or stationary) observer. Consequently, as the source recedes, the light frequency perceived by the observer shifts toward the red (relative to the source) in the visible spectrum, a *redshift*. Conversely, as the source approaches at a relative velocity v, the light frequency shifts toward the blue portion of the visible spectrum, *blueshift*. Therefore, we may write Eq. (5.10) in its more general form as

$$f_O = f_S\sqrt{\frac{1-(\pm v/c)}{1+(\mp v/c)}} \quad (5.11)$$

for zero angle of viewing and applicable to the relative recession (v positive) or relative approach (v negative) of distant light-emitting bodies as perceived by the observer. Thus, it follows that when viewing distant galaxies, the angle of viewing will be essentially zero. Note that the Doppler effect, given by Eqs. (5.10) and (5.11), can be expressed in terms of frequency, wavelength, or energy, and apply equally well to an observer in motion relative to the source. Keep in mind that the velocity, v, is a relative velocity making the Doppler effect symmetric in that respect.

An interesting relationship exists between the Bondi K-factor, Eq. (5.10), and Eq. (3.30a) representing the relativistic flight of a spacecraft that emits a light signal f_S perceived by a stationary observer (relative to the spacecraft) to be f_O. See Problem 6 at the end of this chapter.

5.2.3 Redshift—Velocity and Distance Determinations

The redshift represents a change in the frequency of light in which the frequency received by an observer is less or redder than when it was emitted at the source. This happens as a result of the Doppler effect when the source is receding from the observer as expressed by Eq. (5.10). Astronomers observe a Doppler redshift when observing distance galaxies. This redshift, which occurs when light from the distant galaxies is shifted to the lower frequencies, is evidence that the universe is expanding. On a cosmic scale, it is found from redshift measurements that the distance between galaxies in the universe is increasing at an increasing rate—the increasing expansion rate of the universe. However, the frequency shift effects are more complicated than these statements would imply and involve blueshifts as well as redshifts as explained by the following.

The Doppler redshift can be represented in the slightly different but more useful form as

$$z = \frac{f_S - f_O}{f_O} = \frac{f_S}{f_O} - 1 = \frac{\sqrt{1 - v^2/c^2}}{1 - v/c} - 1. \tag{5.12}$$

This, of course, is the *redshift* observed by our astronomers when they view very distant galaxies receding at a relative velocity v, and is indicative of an expanding universe. If the velocity of recession v is much less than c ($v \ll c$), then Eq. (5.12) takes the approximate classical form

$$z = \frac{\Delta f}{f_O} = \left[\frac{\sqrt{1 - v^2/c^2}}{1 - v/c} - 1\right] = \left[\frac{\sqrt{1 - v^2/c^2} - 1 + v/c}{1 - v/c}\right] \cong \left(\frac{v}{c}\right)_{v \ll c} \tag{5.13}$$

where $\Delta f = f_S - f_O$. The classical form of Eq. (5.13) will be shown later to be useful for redshift calculations of luminescent bodies that are of intermediate distance from the earth (see Hubble's law).

The issue of cosmic light-spectrum shifts is much more complicated and involved than has been implied. Thus, some explanation is needed as to the origin of the redshifts and blueshifts observed by astronomers. It is true that the recession movement of sources such as galaxies creates a redshift effect via Eqs. (5.10) and (5.12). However, a redshift or blueshift may occur from nearby stars that may be receding or approaching the earth, again a Doppler effect given by Eq. (5.11). But it is also true that the outer arms of rotating edge-on galaxies can also create shifts in the light spectrum due to the Doppler effect. Spiral arms on opposite sides of a rotating galaxy produce a redshift if receding from the earth observer or a blueshift if approaching the earth observer (v is negative). Redshifts can also occur due to a dramatic drop-off in the ultraviolet spectrum of hydrogen, known as the "Lyman break," which redshifts into the detectable wavelength range. There appear to be about 50 so-called Lyman-break galaxies. Another origin of redshifts is due to an expansion of space. Light rays "stretch" and undergo a redshift as they travel through expanding space, evidence that the universe is expanding. If the universe were contracting, light rays would be "compressed" and would undergo a blueshift due to the collapsing of space. The redshift of very distant galaxies is believed to be due to the stretching of space and not to a recession of the galaxies per se, though the term 'recession velocity' of such galaxies is often but mistakenly encountered. Finally, small Doppler shifts, both red and blue, can be caused by motion of stars arising from the motion of an associated planet in orbit around the star.

The light-spectrum shifts that are observed by astronomers can be measured because the emission spectra for atoms are distinctive and well known. But some light-spectrum shifts appear to be stronger than others. For example, the more distant galaxies often exhibit larger redshifts than closer ones and are seen as objects as they were further back in time—light from them has taken longer to reach us due to an expanding universe. The largest redshift measured so far corresponds to the state of the universe about 13.7 billion years ago. These redshifts are observed to come from barely detectable objects assumed to be galaxies at the outer edge of the universe sometimes referred to as the *cosmic horizon*—a limit beyond which space may be expanding faster than the velocity of light, c. If this were true, all objects beyond the cosmic horizon would be undetectable by us. What lies beyond this limit of about 14 billion light years from us is presently unknown.

5.2.4 Hubble's Law

In physical cosmology, Hubble's law (after Edwin Hubble and Milton Humason, 1929) states that the redshift in light emitted from distant galaxies is linearly proportional to their distance from the earth. Since redshift z is a function only of the apparent recession velocity, v, Hubble's law can be expressed by

$$v = H_O D \qquad (5.14)$$

where v is the velocity of recession (in km s^{-1}), D is the proper distance that light has traveled from the galaxy to earth (in Mpc, megaparsecs), and H_O is the Hubble parameter having a value of 71 ± 4 km s^{-1} Mpc^{-1}. Thus, knowing the velocity of recession from redshift measurements of say a distant galaxy, its distance can be roughly estimated from Eq. (5.14). Of course the reverse is also true. That is, by knowing distance D, the apparent velocity of recession can be calculated from Eq. (5.14). As future measurements are made, our understanding of the relation between the recession velocity and distance of galaxies, particularly those at the far reaches of the known universe, is certain to further our understanding of Hubble's law and its application. At present, plots of Eq. (5.14) with v as the ordinate show a slight upward bend at the extreme distances given on the abscissa. This is interpreted as an increasing rate at which cosmic space is expanding.

To demonstrate just how small the redshift is, consider the distance of galaxy M31 from the earth, $D = 2.25 \times 10^6$ Lt Yr $= 0.691$ Mpc. Then by Eq. (5.14), its recession velocity is $v \cong 71 \times 0.69 = 49.0$ km s$^{-1} \ll c$, giving an estimated redshift, by Eq. (5.13), of

$$z = (f_S/f_O) - 1 = v/c = 1.63 \times 10^{-4}$$

or

$$f_S/f_O = 1.000163.$$

Then, for light frequencies in the red range we will assume the reference frequency of the source to be $f_S = 4.000000 \times 10^{14}$ Hz, resulting in an observed frequency of

$$f_O = 4.000000 \times 10^{14}/1.000163 = 3.999348 \times 10^{14} \text{ Hz}.$$

Thus, the redshift is measured to be about 1 part in 6100 compared to the source frequency, f_S. Note that all of the above are based on a Hubble parameter $H_O = 71 \pm 4$ km Mps^{-1}, which is thought to be the best value to date and the one we will use in this book. Other values for H_O include $H_O = 72 \pm 8$ km Mps^{-1} with other estimates ranging from 50 to 100 km Mps^{-1}.

Clearly, the light-spectrum shifts, both redshifts and blueshifts, are indeed extremely small effects, and some are even many orders of magnitude smaller than that just demonstrated for the M31 galaxy. Also, their interpretations are not held in agreement by all scientists in the field, and speculation sometimes abounds. Several of the more contested theories and models have been at issue amongst some of the most knowledgeable people in the field. Even so, we can accept their light shift analyses as a first step in our quest for knowledge of the universe in which we live. But obviously, much is left to be learned!

5.2.5 Parallax Method—Distance Measurements

While the redshift method is most useful for ultra deep field distance measurements, the parallax method is useful for stars in our galaxy and for relatively nearby galaxies outside ours. The parallax method is heliocentric in the sense that use is made of the mean radius of the earth's orbit around the sun to determine the greatest parallax (seconds of arc) in the star's directions from the earth during a year. The unit of parallax is parsec, pc, for parallax seconds. The distance to a star or nearby galaxy in parsec (pc) is given by $D = r/\theta_{rad}$, where r is the average radius of the earth's orbit around the sun. We may take the mean distance of the earth from the sun as $r = 1$ AU $= 93.2 \times 10^6$ mi or $r = 93.2 \times 10^6/1.91 \times 10^{10} = 4.87 \times 10^{-6}$ pc. Therefore, the distance from the earth to a distant star or nearby galaxy is calculated to be $D \cong (4.87 \times 10^{-6})(2.06 \times 10^5)/\theta'' \cong 1.00/\theta''$ in pc (parsecs) or $D = 1.00 \times 10^{-3}/\theta''$ in kpc, if the parallax θ'' is measured in seconds of arc. Here, 1 rad $= (180/\pi)(3600) = 2.06 \times 10^5$ seconds of arc. For conversion purposes, 1 kpc $= 3.26 \times 10^3$ light years. Note that the parallax method is sometimes used in connection with redshift measurements and Hubble's law to compare distances of some stars and nearby galaxies.

5.2.6 Composition of Velocities

In Section 2.7 we derived, from first principles, the Lorentz transformation equations (Eqs. (2.22)) from which we obtained the addition (composition) of velocities given by Eq. (2.23) and in generalized form by Eq. (2.26). Now we will use the Minkowski diagram in Fig. 5.2 to accomplish the same goal but by geometric means.

Consider a third observer on line OS in Fig. 5.2 that is moving at a velocity u relative to the rocket observer traveling at a velocity v but has a velocity w relative to the stationary observer on OT. Our goal is to determine the $K(w)$ factor for the observer on OS relative to the stationary observer on OT in terms of the K-factors $K(u)$ and $K(v)$ for observers on OT and OR, respectively. To do this, consider that a light pulse is issued by the observer at rest on OT and is received by the OS observer at Q at a time

$$\tau_2 = K(w)\tau_0$$

where

$$K(w) = \sqrt{\frac{1 + w/c}{1 - w/c}} \qquad (5.15)$$

following the same form as in Eq. (5.8). Now, consider that a light pulse is sent to the rocket observer at N' on OR at time

$$\tau_1 = K(v)\tau_0 \qquad (5.16)$$

where

$$K(v) = \sqrt{\frac{1 + v/c}{1 - v/c}} \qquad (5.17)$$

in agreement with Eq. (5.8). Immediately following receipt of the signal at N', the rocket observer sends a pulse to the observer on OS that is received at Q at a time

$$\tau_2 = K(u)\tau_1 = K(u) \cdot K(v)\tau_0 = K(w)\tau_0$$

where

$$K(u) = \sqrt{\frac{1 + u/c}{1 - u/c}}.$$

Therefore, it follows that

$$K(w) = K(u) \cdot K(v) = \sqrt{\frac{1 + u/c}{1 - u/c}} \cdot \sqrt{\frac{1 + v/c}{1 - v/c}} = \sqrt{\frac{1 + w/c}{1 - w/c}}. \qquad (5.18)$$

Squaring both sides gives

$$\frac{(1 + u/c)(1 + v/c)}{(1 - u/c)(1 - v/c)} = \frac{1 + w/c}{1 - w/c}$$

and simplifying

$$\frac{(c + u)(c + v)}{(c - u)(c - v)} = \frac{c + w}{c - w}.$$

Now, with a little algebra, we solve for w to give the final results

$$w = \frac{c^2(u+v)}{c^2 + uv} = \frac{u+v}{1 + uv/c^2}, \quad (5.19)$$

which is the same as Eq. (2.23).

Equation (5.19) has wide application in summing two velocities. An example, of special interest to us, is the summing of velocities of a two-stage rocket. If the first stage is moving at a velocity v and a second stage is launched from the first stage along the direction of motion at a velocity u, the relativistic sum of the two velocities is then given by Eq. (5.19). For the classical (Newtonian) case, $uv \ll c$ and the sum is simply $(u+v)$. Equation (5.19) can also be applied to multiple staging by summing the velocity of each stage of the rocket configuration with the resultant velocity for the previous two stages, hence always summing velocities two at a time.

Another application of Eq. (5.19) gives the velocity of light propagating in a moving fluid of refractive index n as discussed in Section 2.1. Let the phase velocity of light, c/n, be represented by u, and the velocity of the liquid by v. In this case Eq. (5.19) takes the form

$$w = \frac{(c/n) + v}{1 + (v/cn)} \cdot \left(\frac{1 - (v/cn)}{1 - (v/cn)}\right) = \frac{c/n + v - v/n^2 - v^2/cn}{1 - (v/cn)^2}$$

or

$$w = \frac{c}{n} + v - \frac{v}{n^2} = \frac{c}{n} \pm v\left(1 - \frac{1}{n^2}\right), \quad (5.20)$$

where the quantities $(v/cn)^2$ and v^2/cn are taken as negligible for $v \ll c$. Note that Eq. (5.20) is exactly the empirical results reported in the literature by Fizeau briefly discussed in Section 2.1. Here, the phase velocity, c/n, is the velocity of light in a stationary liquid, where in a vacuum $n = 1$. As indicated in Section 2.1, a moving fluid exerts a "drag" on the propagation of light. Thus, since the liquid can be flowing with and against the propagation of light, the velocity must be represented as $\pm v$ in Eq. (5.20). If the velocity of the flowing liquid is high enough, it can have a significant influence on the velocity of light propagating in that liquid.

5.3 PROBLEMS

1. The velocity of light in flowing water is needed for a given experiment. If the refractive index of water at 20°C is 1.333, by what increments must the flow of water be increased for each increase in light speed of one part in 10^8? Assume that the flow of water is in the direction of light propagation.
   Answer: $v = 2.287n$ ms^{-1} where $n = 0, 1, 2, \ldots$.

2. A three-stage space probe rocket system is moving in matter-free and gravity-free space at a velocity $0.5c$ relative to earth. Let the leading stage be 1, the intermediate stage be 2, and

the trailing stage be booster-rocket 3. If stage 1 is to be launched from stage 2 at a velocity of 0.89c relative to earth, what must be the launch velocity of stage 2 relative to earth? Assume that rocket 3 is discarded after the launch sequence.
Answer: $0.70c$.

3. The star α Vega (apparent visual magnitude 0.14) in the constellation Lyre has a parallax of 0.123 seconds of arc. What is its distance D from the earth in light years?
Answer: $D = 26.5$ Lt Yr.

4. The "Sombrero" galaxy, M104, is viewed in the constellation Virgo. It has a Doppler redshift $z = 2.90 \times 10^{-3}$, or about one part in 345 compared to the source frequency, f_S, in the red range. (a) Calculate the Doppler velocity of apparent recession, v. (b) Estimate its distance from the earth in light years by using Hubble's law.
Answers: (a) $v = 870$ km s^{-1}. (b) $D = 40 \times 10^6$ Lt Yr.

5. The most distant objects (galaxies) are barely detectable at approximately 13.7 billion light years distant from the earth by using the most powerful telescopes available. (a) Calculate their apparent velocity of recession, v, by using Hubble's law. (b) Calculated the expected Doppler redshift, z, from the results of (a). (c) Calculate the expected observed frequency f_O in Hz if the source frequency is taken as $f_S = 4.0 \times 10^{14}$ Hz.
Answers: (a) $v = 0.995c$. (b) $z = 19.0$. (c) $f_O = 2.0 \times 10^{13}$ Hz in the infrared region.

6. The Bondi K-factor, given by Eq. (5.8), has been used in the derivation of the time dilation expression, the composition of velocities, and is found to be functionally related to the relativistic Doppler effect thereby resulting in Eq. (5.10). (a) Now consider a laser-driven rocket moving away from the earth at a uniform earth-measured velocity v where the laser light beam has a frequency f_S. The earth observers perceive this laser beam to have a frequency of f_O. Show that these frequencies are related to the Bondi K-factor by $K = m_{e0}/m_e = f_S/f_O$ where the mass ratio was previously given by Eq. (4.8) if u_e is replaced by c in Eq. (3.30a). (b) Find the relationship between the Doppler redshift $z = \Delta f/f_O$ in Eq. (5.13) and the mass ratio m_{e0}/m_e by using the Bondi K-factor for $v \ll c$.

CHAPTER 6

Other Prospective Transport Systems for Relativistic Space Travel

In this chapter, we will briefly explore only a few of the many possible propulsion systems for which some of the bits and pieces are available to present-day technology. This chapter is needed to tie up loose ends left over from the previous chapters, mainly the means available for achieving relativistic spacecraft velocities by using present and future technology—the future technology being largely speculative. Chemical rocketry must be regarded as a propulsion system of the past.

6.1   NUCLEAR PARTICLE PROPULSION

Nuclear fission and nuclear fusion rockets are examples of nuclear particle propulsion systems. Here, particles can be emitted energetically by an operating fission or fusion reactor but only a fraction of about one percent of their mass is converted into useful propulsive energy. As technology now stands, a fusion rocket is beyond current engineering capability. Sustained controlled fusion has never been demonstrated and prospects for this seem diminishingly small unless there are significant breakthroughs in this technology. On the other hand, fission rockets can be built and are sustainable, but will only be useful as staged rocket configurations. Typical problems include an effective means of directing a collimated beam of the radioactive decay particles to provide thrust, and the creation of effective shielding to protect the spacecraft's occupants. Very large magnets are required for collimating the fission particles and that adds greatly to the overall mass of the rocket. Significant shielding would be necessary to protect any crew that would be present in the spacecraft. Also, the particulate mass of the efflux would be subject to relativistic mass dilation as discussed in Section 4.4. An additional problem involves the formation of a space charge build-up, due to charged particles, resulting in the loss of thrust. In short, nuclear particle propulsion system technology seems not to be ready for interstellar propulsion application in the near future, if ever.

6.2 MATTER/ANTIMATTER PROPULSION

The combination of matter with antimatter offers a means of converting matter into energy at nearly 100% efficiency according to Einstein's energy equation $E = mc^2$. By smashing subatomic particles together at near light velocities, scientists have so far been able to create about one million antihydrogen atoms, which is about 10^{-21} kg or about 10^{-24} metric tons (tonnes) of antimatter. A matter/antimatter rocket propulsion system can be envisioned as a reactor that combines a quantity of matter (for example hydrogen) with an equal amount of antimatter (antihydrogen) with the release of electromagnetic radiation energy according to $E = mc^2$. If this electromagnetic radiation could be collimated and directed from a magnetic nozzle, relativistic rocket velocities could be realized. Since some of the radiation will be absorbed in the walls of the reaction chamber, it is fair to say that an adiabatic condition cannot be achieved but might be approached. Furthermore, the matter/antimatter reaction would also create subatomic particles such as families of muons, neutrinos, and quarks, and radiation such as gamma radiation, all of which have very different properties. The prospect of directionally collimating these particles and gamma radiation would be a daunting task. In any case, hundreds of thousands of tons of antimatter would have to be produced and contained in magnetic tanks, completely separate from matter, and then combined with matter in a reactor chamber under strictly controlled conditions ready for directional collimation. Thus at present, matter/antimatter propulsion remains of considerable interest with known bits and pieces of the technology, but must be viewed as futuristic.

6.3 LASER SAIL PROPULSION

Conventional rocketry relies on rocket-borne fuel to be transformed into useful propulsive power. So such spacecrafts must move mass that is largely fuel, thus greatly limiting the spacecraft's ability to reach distant star/planet systems. However, there exist several important propulsion systems that fall outside the rocket regime. This means that such propulsion systems require no rocket-borne reaction mass to propel the spacecraft and that the propulsion is produced by means other than by an efflux. The laser sail is one of these propulsion systems. Here, sunlight from a giant collector is focused onto a laser which in turn delivers a powerful coherent beam, via a mirror, onto a huge multipart sail. The resulting radiation pressure can then drive the sail to speeds possibly up to 50% light speed, but so over years not days. The size of the solar collector and sail is thousands and hundreds of miles in diameter, respectively, thereby adding significantly to the implementation difficulties involved. In addition to the massive laser infrastructure and the enormous power required, there is the problem that the sail can go only where the laser points and this is likely to be determined only by earth observers due to the finite velocity of light. Clearly, movement of the sail out of a defined straight trajectory poses a serious problem. Also, to slow the sail down to its destination requires that the sail be split into two parts permitting the center rendezvous portion to be slowed down to its destination by reflected laser light from the discarded outer sail. Then to return to earth the outer portion of the rendezvous portion separates from the rotated inner part which is propelled by reflected laser light from the discarded outer part. Then on approaching our solar system the inner part must split into two parts allowing the discarded outer part to slow the inner part once again. Obviously, a very complicated series of operations are required to execute a round trip to a distant star.

6.4 FUSION RAMJET PROPULSION

In 1960, physicist Robert Bussard conceived of a propulsion system that combines the advantages of the efflux-powered rocket with the advantage of a nonreaction mass technology—no rocket-borne fuel required. Fusion ramjet propulsion features a giant magnetic funnel to scoop up interstellar hydrogen as fuel, which is delivered to a reactor where the fusion process converts it to helium with the production of propulsive power. The system would permit the fusion ramjet spacecraft to move most anywhere in the galaxy at some fraction of the speed of light. The down sides to this system are multifaceted. First, there is no known means of fusing atomic hydrogen to form helium, though the process takes place in our sun. Fusion, as has been carried out in our laboratory experiments, makes use of deuterium and tritium both of which are rare in space. Of course, we still do not know what dark matter is even though it may comprise as much as 90% of the matter in the universe. Another problem is that the hydrogen scooped up in space also creates a drag, due to the piling up of hydrogen on the funnel, and this drag opposes the propulsion produced by the efflux resulting from the fusion process. Conceivably, this could bring the fusion ramjet spacecraft to standstill when passing through a relatively high density of hydrogen gas.

6.5 EXOTIC SPACE TRANSPORT AND PROPULSION SYSTEMS

The propulsion systems mentioned above are based, at least in part, on known theory and experimental results. The bits and pieces for these systems are at hand but leaving much of the important work to future technological breakthroughs. Exotic space transport systems involve concepts that are purely speculative since current knowledge has not progressed to any reasonable level of understanding of the underlying physics and engineering involved. Simply put, we really do not know much about interstellar space. Yes, we can bandy about such subjects as the increasingly expanding universe, black holes, wormholes, dark matter, dark energy, the effect of gravity on the fabric of spacetime, superstring theory, M-theory and branes, and a host of other exotic topics. But what do we really know for certain about the nature and laws of physics that govern interstellar space? The fact is that we do not know and probably will not know about these things until we explore and perform experiments in space itself.

The following provides a brief discussion of four of the more commonly discussed space transport systems that may or may not prove to be of practical use in the future. These are systems for which current principles of physics and flight mechanics are not likely to be applicable. There are 20 or more other propulsion/transport systems that have been studied and that have been reported in the literature (see Endnotes for more information on many of these propulsion/transport systems).

6.5.1 Gravity-Controlled Transport

One need not belong to the antigravitic society to wonder if gravity can be controlled or if negative matter and negative gravity exists in the universe. Certainly, if there is such a thing as negative gravity in the universe, there would be more than the usual excitement regarding such a new discovery; particularly, if it could be controlled and/or reversed. This would mean the likelihood of achieving relativistic velocities in a spacecraft without the use of a reaction fuel mass for efflux propulsion. Of course, we will probably not know if negative gravity exists or if antigravity is possible until we explore interstellar space or until we can devise earth-designed or outer-space experiments that would indicate its presence. The prospects for this happening seem poor at the present time. It must be mentioned that such gravity-controlled transport systems sprang from the early speculations of astrophysicist Herman Bondi and physicist Robert Forward.

6.5.2 Transport in Variable Light Speed Media

Here, we refer to the possibility that the velocity of light may greatly exceed the value of 3.00×10^8 m s^{-1} in unusual media such as along or near thin high-density cosmic strings that are thought by some to crisscross the universe as vestiges of the big bang—this according to the work of theoretical physicist J. Magueijo discussed in Section 4.4. If this were true, then journeys to distant star/planet systems would be possible during an astronaut's life-time with reduced mass ratios and a greatly diminished difference between earth- and astronaut-measured flight times. Calculations were made in Section 4.4 to determine the effect that a variation in light speed would have on space travel. Of course, such transport systems are presumably propulsive and require rocket-borne reaction fuel mass together with all the featured problems discussed previously. More importantly, variable light speed transport does *not* mean superluminal transport but it does stretch the acceptable limits of what can or cannot be assumed. Thus, the work of Magueijo must remain as a peculiarity of his imagination; at least until more is understood about interstellar and intergalactic space including the existence of dark matter and dark energy. The work of the NRC scientists, also mentioned in Section 4.4, must also be regarded as a peculiarity which begs for independent corroboration and further study. However, if this were done, it might indeed be an exciting find that could lend some credence to Magueijo's work.

6.5.3 Wormhole Transport and Time Travel

The total collapse of a large stellar object onto itself is thought to create a "singularity" at its center. As a result, the gravitational field around the singularity is predicted to be so intense that below a critical radius, called the *event horizon*, light itself cannot escape thereby defining the region commonly known as a black hole. According to Einstein's general theory of relativity, gravitational forces result from the distortion of spacetime caused by the presence of mass. If two singularities are formed at different locations in space, it is theorized that a wormhole would be formed by their interacting distortions of spacetime which could create a linkage between the singularities. If the wormhole were large enough to accommodate a space vehicle, intrauniverse travel might be possible. If a multiverse exists, singularities in different universes could allow time travel both forward and backward in time.

Obviously, the statements above stretch the limits of reasonable speculation and raise some serious questions. To begin with, it is not certain that singularities exist at the center of black holes. But if they do exist and can distort the spacetime continuum, it is by no means certain that a pair of singularities can produce a wormhole large enough to permit intrauniverse transport. If a wormhole is formed it could be of any size ranging from a Plank length, about 10^{-33} cm, to a thin tube large enough for a space vehicle, or it may not form at all—no one knows. Then there is the problem of a space vehicle venturing too close to the event horizon of a black hole. There would be no escape. Finally, the laws of physics go out the window in any wormhole transport phenomena leaving no room for further discussion at this time. In short, beyond wild speculation, there is little one can say about wormhole transport or time travel other than to admit that much is left to be learned about the physical universe. But then, nothing is impossible until proven to be so.

6.5.4 Warp Drive Transport

As long as we are on a roll in exercising our imagination, we should mention warp drive transport. Here, it is theorized that the presence of a massive spaceship could warp spacetime sufficiently to contract space in front of the spaceship and expand space behind it. Einstein's general theory of relativity permits the manipulation of spacetime in this manner but would require the presence of a negative matter spacecraft. Negative matter reacts to gravitational forces in a manner opposite to that of ordinary matter, which is always attractive. Thus, negative matter would presumably always be repulsive. Negative matter is not antimatter, which has the same gravitational properties as ordinary matter but is of opposite charge including opposite nuclear force charge. Presumably, negative matter would also have its antinegative-matter counterpart. From experimental results on earth, we know that every particle of ordinary matter has its antiparticle counterpart and that particles and antiparticles will annihilate each other if brought into contact. It is not known if negative matter and/or antinegative matter exist in space. Hence, warp drive transport must be regarded as purely speculative.

APPENDIX A

Fundamental Constants, Useful Data, and Unit Conversion Tables

A.1 FUNDAMENTAL CONSTANTS AND USEFUL DATA

Acceleration due to gravity, g, = 9.81 m s^{-2} = 32.17 ft s^{-2}
Earth–sun distance (mean) = 1 AU = 1.50×10^{11} m = 93.2×10^6 mi
$\quad\quad\quad\quad\quad\quad\quad\quad\quad\quad$ = 1.59×10^{-5} Lt Yr (light years)
$\quad\quad\quad\quad\quad\quad\quad\quad\quad\quad$ = 8.35 Lt min (light minutes)
Planck's constant $\hbar = 6.625 \times 10^{-34}$ J s in SI units
Radius of Earth = 6.38×10^6 m = 3963 statute miles
Velocity of light in a vacuum, c, = 2.998×10^8 m s^{-1} = 186,330 mi s^{-1}
Velocity of sound in air at sea level and 0 °C = 331.3 m s^{-1} = 1087 ft s^{-1} = 741 mi h^{-1}

A.2 UNITS OF CONVERSION

Length and Distance:
$\quad$ 1 km = 0.621 mi
$\quad$ 1 mi = 1.61 km = 5280 ft
$\quad$ 1 Lt Yr = 9.46×10^{15} m = 0.307 parsec = 5.88×10^{12} mi
$\quad$ 1 pc (parsec) = 3.26 Lt Yr
$\quad$ 1 kpc (kiloparsec) = 3.26×10^3 Lt Yr = 1.91×10^{10} mi = 3.075×10^{10} km
$\quad$ 1 AU (astronomical unit) = 4.87×10^{-6} pc = 93.2×10^6 mi = 1.585×10^{-5} Lt Yr

Time:
$\quad$ 1 day = 8.64×10^4 s
$\quad$ 1 year = 3.156×10^7 s

Velocity:
$\quad$ 1 m s^{-1} = 3.60 km h^{-1}
$\quad$ 1 mi h^{-1} = 1.610 km h^{-1} = 0.447 m s^{-1} = 1.467 ft s^{-1}
$\quad$ 1 ft s^{-1} = 0.305 m s^{-1}
$\quad$ 1 mi s^{-1} = 5.280×10^3 ft s^{-1} = 1.610 km s^{-1}

Acceleration:
- 1 m s^{-2} = 3.281 ft s^{-2}
- 1 mi s^{-2} = 1.610 km s^{-2}

Energy and Work:
- 1 J = 0.738 ft lb = 3.72 × 10^{-7} hp h = 1 N m = 1 kg m^2 s^{-2}
- 1 kW h = 3.60 × 10^6 J = 1.34 hp h
- 1 EU (energy unit) = 3.15 × 10^{15} J

Power:
- 1 W = 1 J s^{-1}
- 1 hp (U.S.) = 550 ft lb s^{-1} = 746 W
- 1 hp (metric) = 750 W

Specific Impulse:
- 1 N s kg^{-1} = 1 m s^{-1} = 3.281 ft s^{-1} = 0.102 lbf s lbm^{-1}

Thrust to Efflux Power Ratio:
- 1 N kW^{-1} = 1 × 10^{-3} s m^{-1} = 0.168 lbf hp^{-1}

Force:
- 1 N = 1 kg m s^{-2} = 10^5 dyne = 0.225 lb

Mass:
- 1 Atomic mass unit = 1.6605 × 10^{-27} kg
- 1 Metric ton = 1000 kg

Angle:
- 1 rad (radian) = 57.3° = 2.96 × 10^5 second of arc
- 1° = 0.01745 rad

A.3 METRIC (SI) MULTIPLIERS

PREFIX	ABBREVIATION	VALUE
Terra	T	10^{12}
Giga	G	10^{9}
Mega	M	10^{6}
Kilo	k	10^{3}
Hecto	h	10^{2}
Deka	da	10^{1}
Deci	d	10^{-1}
Centi	c	10^{-2}
Milli	m	10^{-3}
Micro	μ	10^{-6}
Nano	n	10^{-9}
Pico	p	10^{-12}
Femto	f	10^{-15}

APPENDIX B
Mathematical Definitions and Identities

B.1 HYPERBOLIC FUNCTIONS

Definitions:

Hyperbolic sine of $x = \sinh(x) = \frac{1}{2}(e^x - e^{-x})$

Hyperbolic cosine of $x = \cosh(x) = \frac{1}{2}(e^x + e^{-x})$

Hyperbolic tangent of $x = \tanh(x) = \dfrac{\sinh x}{\cosh x} = \dfrac{e^x - e^{-x}}{e^x + e^{-x}} = \dfrac{1 - e^{-2x}}{1 + e^{-2x}}$

$\operatorname{csch}(x) = \dfrac{1}{\sinh(x)} \quad \operatorname{sech}(x) = \dfrac{1}{\cosh(x)} \quad \coth(x) = \dfrac{1}{\tanh(x)}$

If $y = \sinh(x)$, then $x = \sinh^{-1}(y)$

If $y = \cosh(x)$, then $x = \cosh^{-1}(y)$

If $y = \tanh(x)$, then $x = \tanh^{-1}(y) = \frac{1}{2}\ell n\left(\dfrac{1+y}{1-y}\right)$

Identities:

$\cosh^2(x) - \sinh^2(x) = 1$
$\operatorname{sech}^2(x) + \tanh^2(x) = 1$

B.2 LOGARITHM IDENTITIES (BASE b)

Identities:

$\log_b M^p = p \log_b M$

$\log_b \sqrt[q]{M} = \frac{1}{q} \log_b M$

$\log_b \frac{1}{M} = -\log_b M$

APPENDIX C
Derivation of The Rocket Equations

C.1 THE PHOTON ROCKET EQUATION

Referring to Eqs. (3.16) and (3.17), the momentum associated with a single photon is given in the *proper system* by $p_{e,photon} = E_{e,photon}/c = m_{e,photon} \cdot c$, where $m_{e,photon}$ is often but improperly called the "relativistic" mass of a photon. A photon has no rest mass but it does have momentum. For N_e photons the photon momentum becomes $c N_e m_{e,photon} = N_e \cdot E_{e,photon}/c = N_e \cdot \hbar/\lambda$.

Continuing in the proper system, consider a rocket propelled at a uniform velocity v_e in matter-free and gravity-free space along a straight trajectory. At a time t_e an amount of reaction fuel mass dm_e is about to be removed from the rocket of mass m_e (essentially all reaction fuel mass) and converted to directed photon propellant out of the rocket's nozzle at 100% efficiency. At a time $t_e + dt_e$ after removal of mass dm_e, the velocity of the rocket becomes $v_e + dv_e$ resulting in a total momentum of $(m_e - dm_e)(v_e + dv_e)$. The change in momentum then becomes

$$dp_e = (m_e - dm_e)(v_e + dv_e) - (m_e v_e - c\, dm_e) = m_e dv_e + (c - v_e)dm_e, \qquad (C.1)$$

where it is understood that $dN_e \hbar/\lambda = dN_e E_{e,photon}/c = c \cdot dN_e m_{e,photon} = c \cdot dm_e$ indicating that an amount of reaction mass dm_e is converted to useful photon momentum and that the double derivative term $dm_e dv_e$ has been neglected since in the limit of infinitesimals $dm_e dv_e$ is zero.

The external force on the rocket must be equal to the time rate of change of Eq. (C.1) given by

$$F_{Ext} = \frac{dp_e}{dt_e} = m_e \frac{dv_e}{dt_e} + (c - v_e)\frac{dm_e}{dt_e}. \qquad (C.2)$$

But $F_{Ext} = 0$ in the absence of other forces (e.g., gravitational, air resistance, etc.). Therefore,

$$m_e \frac{dv_e}{dt_e} = -(c - v_e)\frac{dm_e}{dt_e}$$

or

$$dv_e = -(c - v_e)\frac{dm_e}{m_e} = -V_{Rel}\frac{dm_e}{m_e}, \qquad (C.3)$$

where $V_{Rel} = (c - v_e)$ is the relative velocity of the photon propellant. But the velocity of the photon efflux is constant at c independent of the velocity of the rocket. Therefore, $V_{Rel} = c$ giving

$$dv_e = -c \frac{dm_e}{m_e}. \qquad (C.4)$$

Assuming that a space flight begins at $v_e = 0$ and that the initial (rest) mass of the rocket is m_{e0}, integration of Eq. (C.4) gives

$$\frac{v_e}{c} = -\ell n \frac{m_e}{m_{e0}} = \ell n \frac{m_{e0}}{m_e}. \qquad (C.5)$$

Equation (C.5) is the *idealized photon rocket equation* given with respect to the proper Einstein number, v_e/c. This equation is to be compared to the relativistic rocket equation, Eq. (3.30), given with respect to the Einstein number v/c and an assumed particulate propellant of velocity u_e. Equating Eq. (C.5) to Eq. (3.30) with u_e replaced by c yields Eq. (3.32), as it must.

C.2 THE EFFECT OF EFFICIENCY

The effect of efficiency should be considered when deriving the rocket equation. In the case of the photon rocket, the efficiency η of converting an amount of fuel mass dm_e to directed photon propellant must affect the $c \cdot dm_e$ term in Eq. (C.1) as $\eta(c \cdot dm_e)$. This means that only a fraction η of the amount dm_e removed from the rocket reaction mass can be converted to useful photon propellant. Following this through the derivation just given results in changing Eq. (C.5) in the following way:

$$\frac{v_e}{c} = -\ell n \left(\frac{m_e}{m_{e0}}\right)^{\eta} = \ell n \left(\frac{m_{e0}}{m_e}\right)^{\eta}. \qquad (C.6)$$

As a result of (C.6), all of the calculations in Chapters 3 and 4 will be affected. In general, this will make journeys to distant stars and galaxies more costly in terms of fuel mass and energy requirements. Calculations with efficiency effects have not been made in this text since η cannot be properly quantified. This is the main reason why the calculations made in Chapters 3 and 4 must be regarded as idealized.

C.3 THE CLASSICAL ROCKET EQUATION

Equations (C.1) through (C.3) apply equally well to a particulate propulsive efflux having a relative velocity $V_{Rel} = (u_e - v_e)$ where $u_e \ll c$ if c is replaced by u_e. Here, as with c, u_e is the effective efflux velocity observed from the rocket's frame of reference whose creation is responsible for the thrust that propels the rocket. Thus, in the proper system, the rocket equation

DERIVATION OF THE ROCKET EQUATIONS

is expressed in terms of the effective efflux velocity u_e (the velocity relative to the rocket) as

$$\frac{v_e}{u_e} = -\ln\frac{m_e}{m_{e0}} = \ln\frac{m_{e0}}{m_e}. \tag{C.7}$$

Equation (C.7) is the *idealized classical rocket equation* widely cited in the field of rocket mechanics and presented in this text as Eq. (3.9). Note that Eq. (C.7) is the same as Eq. (C.5) when u_e is replaced by c. This is not to say that $u_e = c$ but rather it indicates that the derivation of the rocket equation, for either a photon or particulate efflux, is independent of the propulsion system used. Note that Eqs. (C.7) must be altered as in (C.6) if efficiency is to be taken into account.

Taking the antilog of Eq. (C.7) there results

$$\frac{m_e}{m_{e0}} = e^{-v_e/u_e}. \tag{C.8}$$

When Eq. (C.8) is directly introduced into Eq. (3.31) the result is Eq. (3.32) which is another equation found frequently in the literature relating velocities observed from the point of view of two frames of reference, one on earth and the other by astronauts in the rocket. See Section 3.4 for details.

Endnotes

Note. The references cited below provide a few useful sources of information selectively taken from the vast supply that is currently available.

A somewhat outdated but useful introduction to rocket mechanics is presented by E. M. Goodger, *Principles of Spaceflight Propulsion*, International Series of Monographs in Aeronautics and Astronautics, Pergamon Press, Oxford, 1970.

One of the first attempts to apply Einstein's special theory of relativity to rocket spaceflight is presented by Eugen Sanger, "Flight Mechanics of Photon Rockets," Aero Digest, July 1956, pp. 68–73. (Based on a paper presented at the International Conference on Jet Propulsion at Freudenstadt, Germany, February 8, 1956.)

A plethora of information on the basics of spaceflight can conveniently be obtained on the JPL/NASA website http://www2.jpl.nasa.gov/basics/guide.html

An easy read for those wanting to know more about Einstein's special and general theories of relativity is found in the paperback by book George Gamow: *Biography of Physics*, Harper Torchbooks, Harper & Row, New York, NY, 1961. As always, Gamow is both informative and entertaining.

The best source of coverage for the Special Theory of Relativity and the relativistic Doppler effect is Einstein's original paper "On the Electrodynamics of Moving Bodies." See *On the Shoulders of Giants, The Great Works of Physics and Astronomy*, Ed. Stephen Hawking, Running Press, Philadelphia, PA, 2002.

A reasonably good presentation of special theory and the use of Minkowski diagrams is presented by David Bohm, *The Special Theory of Relativity*, Addison-Wesley (Advanced Book Classics), New York, NY, 1989. This book was first published from Dr. Bohm's lecture notes by W. A. Benjamin in 1965.

An article that breaks with traditional thought on the speed of light is that of Andreas Albrecht and João Magueijo, "Time Varying Speed of Light as a Solution to the Cosmological Problems," *Physical Review D,* **59**, pp. 043516 1-13 (1999).

For a detailed derivation of the relativistic rocket equation that parallels the derivation used in this text, see *A transparent Derivation of the Relativistic Rocket Equation*, by Robert L. Forward in the 31st AIAA/ASME/SAE, ASEE Joint Propulsion Conference and Exhibit, July 10–12, 1995, San Diego, CA.

The web version of the Usenet Physics FAQ (Frequently Asked Questions) is a valuable source of information on a variety of subjects that relate to those in this text. The articles in this FAQ are based on discussions and information from good reference sources. The articles include general physics, special relativity, general relativity, and cosmology, black holes, etc. Enter http://math.ucr.edu/home/baez/physics/ or simply enter and search *Usenet Physics FAQ*.

A modern treatment of present-day thinking is elegantly presented by Brian Green, *The Elegant Universe*, W. W. Norton & Company, New York, 1999. Green's book can be digested by most educated persons.

The lay person may find the following book both useful and entertaining: Stephen Hawking, *The Universe in a Nutshell*, Bantam Books, New York, NY, 2001. The artwork is outstanding.

An updated account of star/planet systems can be found on the web: http://planetquest1.jpl.nasa.gov/atlas_search.cfm and http://zebu.uoregon.edu/newplanet.html

Index

Acceleration, xi, 33–35, 48, 49
   constant, 35, 59, 61–70, 73, 76
   due to gravity, 35, 37, 101
   proper, 33–35, 49, 59–63
   variable, 34, 74–76
Accelerometer, xi, 33
Addition of velocities, 20, 23–25, 91–93
Adiabatic condition, 36, 96
   process, xi, 77
   propulsion system, 77
Antigravity, xi, 98
Antimatter, xi, 29, 96
Antiparticle, xi, 100
Astronomical unit (AU), 91, 101
Average proper Einstein number, 61–70, 73–76
Black hole, xi, 99
Blueshift, xi, 88, 89
Bondi, Hermann, xi, 84
Bondi K-factor, 84
Burnout mass, xi, 36, 78
Calculation plan, xi, 62–64
Characteristic value, xi, (see Proper quantity)
Chronometer, 33, 34, 35
Classical rocket equation, 36, 108–109
Clock paradox, xii, 15–17
Composition of velocities, 20
Conservation of energy, xii, 30
Conservation of momentum, xii, 25, 26, 107
Coordinates
   transformation of, 20–23
   in Minkowski diagrams, 83, 84

Correspondence principle, xii, 35, 46
Cosmic horizon, xii, 72, 73, 90
Cosmic strings, xii, 72, 98
Cosmology, xii, 72, 90
Dark energy, xii, 1, 2, 27, 72, 80, 97, 98
Dark matter, xii, 1, 2, 27, 72, 97, 98
Deuterium, xii, 97
Doppler effect, xii, 85, 87–89
   blueshift, xi, 88, 89
   redshift, xviii, 85, 88, 89, 94
   relativistic, xviii, 85, 87, 94
Dynamics, xii, 1,
Eigenvalue, (see Proper quantity)
Einstein, Albert, 9, 10
Einstein number, xiii, 44
Efficiency
   considerations, 77
   conversion, 56, 62, 66, 77, 96
   frozen flow, xiv, 39, 43
   internal, xv, 39, 43, 77
   propulsive, xviii, 37, 38, 42, 43
   thermal, xix, 29, 43, 77
   thrust, xix, 39, 43
   total, 39, 43, 77, 82
Efflux, xiii
   particulate, 39, 70–71
   photon, 39–44, 62, 77
   power, xiii, 37, 38, 42–44, 71
   residual power, 38, 42
   velocity, 21, 24, 33, 35–39, 43, 70–71

Electromagnetic radiation, xiii, 5, 29
   frequency, 5
   wavelength, 5
Electromagnetic spectrum, 5, 77, 88, 89, 91
Electron, xiii
Elementary particle, xiii
Energy
   kinetic, 28, 29
   requirements, 54–55
   rest, xix, 28
   specific, xix, 54–55, 78–79
   total, 28, 79–80
Escape velocity, xiii, 2
Ether wind, xiii, 6, 8–9
Event, xiii
Event horizon, xiii, 99
Exotic space transport systems, 97–98
Fitzgerald, G. F., 9, 17
Fizeau, Armond, 6, 93
Flight plan for space travel, xiii, 60–62
Force, xiii, 27, 35, 107
Free space, xiii, 33, 38, 56, 62, 63, 96
Frequency, xiii, 5
Frozen flow efficiency, xiv, 39, 43
Fundamental constants and useful data, 101, 103
Fusion ramjet propulsion, xiv, 97
Galaxy, xiv, 68–69
Galilean transformation equations, xiv, 20–22
Galilei, Galileo, 20
Gamma rays (radiation), xiv, 5, 77, 80–81, 96
General relativity, xiv, 99
Gravitational force field, xix
Graviton, xiv
Gravity controlled transport, 98
Hubble, Edwin, 90
Hubble parameter, xiv, 90–91

Hubble's law, xiv, 90–91
Hyperbolic functions, 105
Imaginary time, Minkwoski diagrams, xiv, 83–84
Index of refraction, 6, 93
Inertial frame of reference, xv, 10, 11, 13, 21, 25
Internal efficiency, xv, 39, 43, 77
K-calculus and relativistic effects, 85–91
K-factor, 84, 86–88
Kinematics, xv, 1
Kinetic power, xv, 37, 42
Laser sail propulsion, xv, 96–97
Life preserving issues, 80–81
Light cone, xv, 84
Light, nature and velocity of, 5–6, 10
Light year, xv, 101
Logarithm identities, 105
Lorentz boost, 33
Lorentz contraction, xv, 16, 18, 19
Lorentz, Hendrik, 9
Lorentz transformation equations, xv, 20
Lyman break, xv, 89
Mach number, xv, 40
Magueijo, João, 72, 98
Mass-energy expression, 28
Mass, xvi
   burnout, 36, 78
   initial, 36, 40
   instantaneous, 36, 40
Mass ratio, xvi, 36, 40, 77–78
Mathematical identities, 105
Matter/Antimatter propulsion, 96
Metric multipliers, 103
Michelson, Albert, 6
Michelson-Morley experiment, 6, 7–9
Minkowski diagrams, xvi, 83–87, 91

Minkowski, Herman, 83
Multiverse, xvi, 99
Muon, xvi, 16, 17, 77, 81, 96
Nature of light, 5
Negative matter, xvi, 98, 99–100
Neutrino, xvi, 77, 81, 96
Newton, Isaac, 10
Newton's laws of motion, xvi, 10, 27, 35, 41
Noninertial frame of reference, xvii, 11, 13, 33–34
Nuclear particle propulsion, xvii, 95
Parallax method, xvii, 91
Parsec, distance measurement, xvii, 91, 101
Particulate propulsion, xvii, 79–71
Payload, xvii, 77–78
Permeability in a vacuum, 5
Permittivity in a vacuum, 5
Phase velocity, xvii, 93
Photoelectric effect, xvii, 10
Photonic rocket propulsion, xvii, 39–44
Photonic thrust, 43, 44
Photon rocket equation, 40, 107–108
Pion, 29
Plank length, xvii, 99
Plank's constant, xvii, 40, 41
Positron, xviii, 29
Postulates of special relativity, xviii, 10–11
Practical considerations for space flight, 76–81
Propellant, xviii
Proper Einstein number, xviii, 44
Proper quantity (eigenvalue), xviii, 13
Propulsion
    particulate vs. photonic, 70–71
Propulsive efficiency, xviii
    particulate rocket, 37
    photon rocket, 42

Quark, xviii, 77, 81, 96
Reaction mass, xviii, 96, 97
Redshift, xviii, 88–90
Relativistic effects, xviii
Relativistic rocket equation, 46
Relativistic transformation equations, xix, 23
Requirements for interstellar space travel, 3
Rest energy, xix, 28
Rest frame of reference, xix, 34
Rest mass, xix, 26, 28–30
Rocket equation, xix
    classical, 36, 108–109
    relativistic, 46
    photon, 40, 108–109
Simultaneity, 11–13
Singularity, xix, 99
Spacetime, xix
Special theory of relativity, xix, 10
Specific energy, xix, 54–55, 78–79
Specific impulse, xix, 37, 41–42
Star/Planet systems, 64, 65, 112
Superluminal, xix, 72
Synchronicity, 11–13
Tachyons, xix, 27
Thermal efficiency, xix, 29, 43, 77
Thrust, xx, 36–39, 41–44, 60
Thrust efficiency, xx, 39, 43
Thrust-to-efflux power ratio, 39, 43
Time dilation, xx, 13–17, 85–87
Time travel, 99
Total efficiency, 39, 43, 77, 82
Tritium, xx, 97
Tsiolkovsky, Konstantin, 36
Tsiolkovsky rocket equation, xx, 36, 108–109
Twins paradox, 15–17
Units of conversion, 101–102
Variable light speed transport, 98

Varying proper acceleration, 74–77
Varying speed of light, 71–74, 76
Velocity of light,
   in Bose-Einstein condensate, 72
   in ether wind, 8–9
   in liquid media, 71
   in vacuum, 5
   in water, 6, 93
Warp drive transport, xx, 99
Warp number, xx, 40
Wavelength, xx, 5, 88, 107
Wormhole, xx, 99
Wormhole transport, 99

Biography

Professor Tinder's teaching interests have been highly variable over his tenure at Washington State University (WSU). They have included crystallography, thermodynamics of solids (both equilibrium and irreversible thermodyamics), tensor properties of crystals, advanced dislocation theory, solid state direct energy conversion (mainly solar cell theory, thermoelectric effects, and fuel cells), general materials science, advanced reaction kinetics in solids, electromagnetics, and analog and digital circuit theory. In recent years he has taught logic design at the entry, intermediate and advanced levels and has published a major text in the area. He has conducted research and published in the areas of tensor properties of solids, surface physics, shock dynamics of solids, milli-micro plastic flow in metallic single crystals, high speed asynchronous (clock independent) state machine design, and Boolean algebra (specifically XOR algebra and graphics).

Professor Tinder holds a B.S., M.S. and Ph.D. all from the University of California, Berkeley. It was there, as a graduate student, he gave his first graduate seminar on relativistic rocket mechanics and space travel to a variety of students and professors in engineering and the physical sciences. He has since lectured on the subject to a discussion group of professors at WSU. Over the years he has cultivated an increasingly active interest in this area while gathering more information from a variety of sources. It is this information together with his own work that Professor Tinder presents in the contents of this text.

He has spent one year as a visiting faculty member at the University of California, Davis, in what was then the Department of Mechanical Engineering and Materials Science. Currently, he is Professor Emeritus of the School of Electrical Engineering and Computer Science at WSU where he has been a major contributor to the Computer Engineering program there over a period of two decades.

Printed in the United States
153039LV00003B/190/P